MACAT

An Analysis of

Charles Darwin's

On the Origin of Species

by Means of Natural Selection,
or The Preservation of Favoured
Races in the Struggle for Life

Kathleen Bryson
and
Nadejda Josephine Msindai

ROUTLEDGE

Published by Macat International Ltd
24:13 Coda Centre, 189 Munster Road, London SW6 6AW.

Distributed exclusively by Routledge
2 Park Square, Milton Park, Abingdon, Oxon OX14 4RN
711 Third Avenue, New York, NY 10017, USA

Routledge is an imprint of the Taylor & Francis Group, an informa business

www.macat.com
info@macat.com

Cataloguing in Publication Data
A catalogue record for this book is available from the British Library.
Library of Congress Cataloguing-in-Publication Data is available upon request.
Cover illustration: Capucine Deslouis

ISBN 978-1-912302-36-9 (hardback)
ISBN 978-1-912128-63-1 (paperback)
ISBN 978-1-912281-24-4 (e-book)

CONTENTS

THE MACAT LIBRARY

The Macat Library is a series of unique academic explorations of seminal works in the humanities and social sciences – books and papers that have had a significant and widely recognised impact on their disciplines. It has been created to serve as much more than just a summary of what lies between the covers of a great book. It illuminates and explores the influences on, ideas of, and impact of that book. Our goal is to offer a learning resource that encourages critical thinking and fosters a better, deeper understanding of important ideas.

Each publication is divided into three Sections: Influences, Ideas, and Impact. Each Section has four Modules. These explore every important facet of the work, and the responses to it.

This Section-Module structure makes a Macat Library book easy to use, but it has another important feature. Because each Macat book is written to the same format, it is possible (and encouraged!) to cross-reference multiple Macat books along the same lines of inquiry or research. This allows the reader to open up interesting interdisciplinary pathways.

To further aid your reading, lists of glossary terms and people mentioned are included at the end of this book (these are indicated by an asterisk [*] throughout) – as well as a list of works cited.

Macat has worked with the University of Cambridge to identify the elements of critical thinking and understand the ways in which six different skills combine to enable effective thinking.
Three allow us to fully understand a problem; three more give us the tools to solve it. Together, these six skills make up the **PACIER** model of critical thinking. They are:

ANALYSIS – understanding how an argument is built
EVALUATION – exploring the strengths and weaknesses of an argument
INTERPRETATION – understanding issues of meaning

CREATIVE THINKING – coming up with new ideas and fresh connections
PROBLEM-SOLVING – producing strong solutions
REASONING – creating strong arguments

To find out more, visit **WWW.MACAT.COM.**

CRITICAL THINKING AND *ON THE ORIGIN OF SPECIES BY MEANS OF NATURAL SELECTION*

Primary critical thinking skill: CREATIVE THINKING
Secondary critical thinking skill: PROBLEM-SOLVING

Charles Darwin called on a broad and unusually powerful combination of critical thinking skills to create his wide-ranging explanation for biological variety, *On the Origin of Species By Means of Natural Selection.*

It's one of those rare books that takes a huge problem—the enormous diversity of different species—and seeks to use a vast range of evidence to solve it. But it was perhaps Darwin's towering creative prowess that made the most telling contribution to this masterpiece, for it was this that enabled him to make the necessary fresh connections between so much disparate evidence from such a diversity of fields.

All of Darwin's critical thinking skills were required, however, in the course of the decades of work that went into this volume. Taken as a whole, Darwin's solution to the problem that he set himself—how to explain speciation—is carefully researched, considers multiple explanations, and justifies its conclusions with well-organised reasoning. At the time of the publication, in 1859, there were various explanations for the changes that Darwin—and others—observed; what separated Darwin from so many of his contemporaries is that he deployed critical thinking to arrive at a significantly new way of fitting explanation to evidence; one that remains elegant, complete and predictive to this day.

ABOUT THE AUTHOR OF THE ORIGINAL WORK

Charles Darwin was born in 1809 in Shrewsbury, England. He was a member of a wealthy and well-connected family, which included many physicians and scientists. Darwin graduated from Christ's College, Cambridge, in 1831, and a few months later took a position as ship's naturalist on board HMS *Beagle*, a Royal Navy ship then embarking on a surveying voyage to South America.

The voyage proved to be a turning point in Darwin's life. Using the information he collected along the way, he was able to formulate a theory to explain how evolution works. He called this 'natural selection.' Darwin published *On the Origin of Species by Means of Natural Selection* in 1859 and went on to write another 25 books. He died in 1882 and is buried in Westminster Abbey, London.

ABOUT THE AUTHORS OF THE ANALYSIS

Kathleen Bryson has a PhD in evolutionary anthropology from University College, London. The focus of her research is on the cognitive and adaptative roots of prejudice and discrimination in humans and other apes.

Nadezda Josephine Msindai is a PhD candidate in evolutionary anthropology at University College, London, where she is examining construction behaviour in an introduced population of chimpanzees.

ABOUT MACAT

GREAT WORKS FOR CRITICAL THINKING

Macat is focused on making the ideas of the world's great thinkers accessible and comprehensible to everybody, everywhere, in ways that promote the development of enhanced critical thinking skills.

It works with leading academics from the world's top universities to produce new analyses that focus on the ideas and the impact of the most influential works ever written across a wide variety of academic disciplines. Each of the works that sit at the heart of its growing library is an enduring example of great thinking. But by setting them in context – and looking at the influences that shaped their authors, as well as the responses they provoked – Macat encourages readers to look at these classics and game-changers with fresh eyes. Readers learn to think, engage and challenge their ideas, rather than simply accepting them.

'Macat offers an amazing first-of-its-kind tool for interdisciplinary learning and research. Its focus on works that transformed their disciplines and its rigorous approach, drawing on the world's leading experts and educational institutions, opens up a world-class education to anyone.'

Andreas Schleicher
Director for Education and Skills, Organisation for Economic Co-operation and Development

'Macat is taking on some of the major challenges in university education … They have drawn together a strong team of active academics who are producing teaching materials that are novel in the breadth of their approach.'

Prof Lord Broers,
former Vice-Chancellor of the University of Cambridge

'The Macat vision is exceptionally exciting. It focuses upon new modes of learning which analyse and explain seminal texts which have profoundly influenced world thinking and so social and economic development. It promotes the kind of critical thinking which is essential for any society and economy. This is the learning of the future.'

Rt Hon Charles Clarke, former UK Secretary of State for Education

'The Macat analyses provide immediate access to the critical conversation surrounding the books that have shaped their respective discipline, which will make them an invaluable resource to all of those, students and teachers, working in the field.'

Professor William Tronzo, University of California at San Diego

WAYS IN TO THE TEXT

KEY POINTS

- **Charles Robert Darwin (1809–82) was a British naturalist* (a scholar of the natural world) and geologist* (a scholar of the formation and history of the earth's rocks).**

- *On the Origin of Species by Means of Natural Selection* **explains the origin, succession, and extinction of species by "natural selection"* (the process by which organisms over-reproduce and then compete for resources and survival; those best adapted to a particular environment live to repoduce and thus successfully pass on their useful adaptations to the next generation).**

- **The main concepts in** *On the Origin of Species* **are central to the life sciences.**

Who Was Charles Darwin?

Charles Darwin, the author of *On the Origin of Species by Means of Natural Selection* (1859), a work explaining evolution* by the process he called "natural selection," was a renowned natural historian.* His extraordinarily influential book changed the way scientists understand the origin and development of life on earth, and earned its author the title "father" of evolutionary theory.

Darwin was born in 1809 in the market town of Shrewsbury in England. As part of the well-known Darwin-Wedgwood family, he was born into both wealth and privilege. In 1831, he graduated from Christ's College, Cambridge, with a BA; the following winter he became ship's naturalist aboard the HMS *Beagle*,* a Royal Navy vessel on a five-year scientific trip around the globe. On Darwin's return, his father organized investments that allowed him to be a "self-funded gentleman scientist."[1]

Darwin's social circle was one of scientific academics and radical thinkers such as the geologist Charles Lyell* and the social theorist Harriet Martineau,* the former of whom helped him to develop his theory of natural selection. Darwin initially delayed publication of his theory, fearing the public outrage that would follow; this was because it seemed to refute Christian teachings regarding the origins of life on earth (among other things). In 1858, however, the naturalist Alfred Russel Wallace* (1823–1913), who was also a collector and tropical field biologist (a specialist in tropical organisms), wrote to Darwin proposing a similar theory for the mechanism of evolution.This forced Darwin's hand, and he decided to publish immediately.

In 1858, Darwin and Wallace's mutual theory of evolution by natural selection was published in an article presented to the Linnean Society* (an institution founded in London for the discussion and study of natural history).The following year, Darwin published a more detailed account of his theory in *On the Origin of Species*. In total, he published 25 books on diverse topics such as barnacles, plants, and earthworms. Having devoted his life to scientific research, he died in the county of Kent at his family home, Down House, in April 1882. He is buried at Westminster Abbey, in London.[2]

What Does *On the Origin of Species* Say?

The main subject of *On the Origin of Species* is evolution—the process by which all earth's species have descended from a common ancestor. The theory of natural selection, explaining how organisms adapt to their environment to ensure their survival, is based on the observation that while plants and creatures tend to over-reproduce, resources are finite, yet population sizes generally remain stable.

To account for this, Darwin points out that there are fine variations between organisms, notably in their hardwired behavior (what he calls "instinct"). As a species is made up of more individuals than can possibly survive all together, there is a struggle between them for existence. In this competition for limited resources, those individuals best adapted to their environment are more likely to survive and pass on their traits (the differences in their behavior and physical attributes). Gradually, a species evolves to occupy a different niche, so becomes a new species, and is unable to reproduce with the species from which it descended.

The concept of natural selection remains a major part of the study of evolution, along with mutation* (a relatively sudden change in an animal's genetic makeup, leading to a change in its behavior or physical constitution), and genetic drift* (roughly, the process by which genetic information changes over time as animals reproduce).

Secondary themes in *On the Origin of Species* include sexual selection,* speciation* (the process by which a new species evolves), and gradualism* (the way in which species change in intermediate stages over time).

Sexual selection (natural selection via the choice of a mate, for example) can take place either within the sexes or between them. In intrasexual selection,* members of the same sex compete for mates of the opposite sex. This is mostly seen in competition between males: the male with the best fighting technique or largest body size will have the highest chance of winning. The winner then gains exclusive access to mates, and so out-reproduces the losers; natural selection occurs if the characteristics that determine the outcome of the contest are inherited. In intersexual selection, sometimes referred to as "female choice," members of the sex with the more expensive gametes (usually females) choose among potential mates based on certain qualities. If the preferred individuals are genetically different from their rivals, natural selection occurs.

Darwin sees evolution as a slow and gradual process. In *On the Origin of Species*, he introduces the concept of gradualism—species evolving and accumulating small variations over long periods of time. He explains, too, how a new species can arise by way of population speciation, which occurs when a population divides and two new species develop. If a species occupies a large area, variations in the environment will mean that various individuals experience different pressures, and so adapt differently.

Within separate regions, new species arise because natural selection acts independently in each environment. Darwin suggests that there must also be hybrid zones between regions; the individuals within them, being less well adapted to the conditions of either adjoining zone, eventually become extinct.

Why Does *On the Origin of Species* Matter?

On the Origin of Species is one of the most influential scientific texts ever written. The idea of natural selection proposed within it provided an explanation both for the evolution of complex organisms and the exquisite fit of those organisms to their environment, without the need for divine intervention. Darwin used extensive evidence to support his theory of descent with modification*—the pattern of evolution—and the idea that all living organisms are related by way of a common ancestor.

Perhaps even more remarkable was that Darwin wrote the book in such a way that both scientists and the general public could understand his theory. This wide readership resulted in controversy, however, as the text was regarded as a challenge to the prevailing ideology of state and Church.

In establishing the theory of evolution, Darwin laid the foundations for the field of evolutionary biology.* To a lesser extent, he had an influence on disciplines as diverse as modern philosophy, oceanography* (the scientific study of oceans), linguistics* (the study of the structures and nature of language), sociocultural anthropology* (the study of human cultural and social life), history, and economics.

Nearly 160 years after the publication of *On the Origin of Species*, however, evolution is still a contentious issue. Many remain skeptical, and fewer than half (48 percent) the people in the United States accept Darwin's theory of natural selection.[3] At the time, atheists* (those who do not believe in any god) and materialists* (those who look for material causes alone for all physical phenomena) thought it represented a serious, decisive challenge to religion. Evolution and natural selection justified dismissing the idea that a deity may have played a role in creating humans and animals.

Religious believers, however, who saw the wonders of the natural world as God's creation, regarded the theory as a great insult. Referencing positivism,* the principle that knowledge can only advance through scientifically verifiable facts, the evolutionary biologist Ernst Mayr* wrote: "Eliminating God from science made room for strictly scientific explanations of all natural phenomena; it gave rise to positivism, and produced a powerful intellectual and spiritual revolution, the effects of which have lasted to this day."[4] Even today, religious sentiment is mixed. Some accept the concept of natural selection but others still oppose Darwin's theory.

While there have been many developments in evolutionary theory since the publication of *On the Origin of Species*, the theory of natural selection still holds as a mechanism for explaining species change;[5] and Darwin's ideas have come to influence strongly both science and society.

NOTES

1   Janet Browne, *Charles Darwin: A Biography. Vol. 1: Voyaging* (Princeton: Princeton University Press, 1996), 434–5.

2   Janet Browne, *Charles Darwin: A Biography. Vol. 2: The Power of Place* (London: Pimlico, 2003), 496.

3   Pew Forum, "Religious Groups: Opinions of Evolution," February 4, 2009. Available at http://www.pewforum.org/2009/02/04/religious-differences-on-the-question-of-evolution/. [Accessed February 5, 2016,]

4   Ernst Mayr, "Darwin's Influence on Modern Thought," *Proceedings of the American Philosophical Society* 139, no. 4 (Dec 1995): 317–25, *Scientific American*, July (2000), 79–83.

5   H. Allen Orr, "Testing Natural Selection with Genetics," *Scientific American* 300, no. 1 (2009): 44.

SECTION 1
INFLUENCES

MODULE 1
THE AUTHOR AND THE
HISTORICAL CONTEXT

KEY POINTS

- Evolutionary* theory and the mechanism of natural selection together provide the framework for all contemporary biological sciences.

- Charles Darwin's social standing and his proximity to the leading intellectual minds of his time facilitated the development of his theory.

- Mid-nineteenth-century Europe was bursting with post-Enlightenment* reformist ideas concerning slavery, colonialism,* and the nature of man. (The Enlightenment emphasized notions of rationality and liberty; colonialism is the process by which one nation or people exploit another through the occupation of land.)

Why Read This Text?

Charles Darwin's *On the Origin of Species by Means of Natural Selection* (1859) explains how species have evolved from a common ancestor through the process of natural selection. Darwin drew on different sources to formulate a theory that would revolutionize science, at a time when the advancement of knowledge was restricted by religious dogma. Afterward, evolution became a legitimate field of scientific inquiry.[1]

The text is relevant to studies of humans, animals, and plants,[2] offering a theory that still serves as a valid explanation for species change.[3] It is central to studies such as biology, evolutionary psychology* (the study of the human mind and behavior conducted

> **66** Is man an ape or an Angel? I, my Lord, am on the side of the angels. I repudiate with indignation and abhorrence those new-fangled theories.**99**
>
> British Prime Minister Benjamin Disraeli, speech, 1864, quoted in *Charles Darwin: The Man and His Influence* by Peter Bowler

in the light of evolutionary theory), and botany* (the study of plants). To a lesser extent, it also influenced modern philosophy, oceanography (the scientific study of oceans), linguistics (the study of the structures and nature of language), sociocultural anthropology (the study of human cultural and social life), history, and economics.

Author's Life

Charles Robert Darwin was born in 1809 to an affluent, intellectual family. In an effort to give him a respectable occupation, his father sent him to the University of Edinburgh Medical School at the age of 16. Darwin had no desire to be a doctor. He found the curriculum dull and the sight of operations performed without anesthetic disturbing. Instead, he became interested in natural history (the natural world), reading widely, collecting specimens, and dissecting small animals. He befriended a taxidermist, whpooro taught him how to skin and stuff birds, and the zoologist* Robert Grant,* who introduced him to the French biologist Jean-Baptiste Lamarck's* theory of transmutation* (an early attempt to explain the idea that changes can be passed from generation to generation).[4] He joined the Plinian Society*—a club for students interested in natural history—and took part in debates on science.

Darwin's neglect of his medical studies annoyed his father and so, in 1827, he was sent to Christ's College, Cambridge, to study theology. But Darwin's interest in natural history continued. He took lectures in botany and went on long plant-collecting expeditions with the clergyman and botanist John Stevens Henslow.* Darwin also developed an interest in geology* (inquiry into the formation and history of the earth's rocks) and attended a course on the subject taught by the geologist Adam Sedgwick.*

In 1831, after completing his studies at Cambridge, Darwin took up a post as ship's naturalist and gentleman companion to Captain Robert Fitzroy* on board the HMS *Beagle*. Darwin had to pay his own way and the trip lasted longer than he had expected—five years instead of two. During the voyage, he collected numerous specimens, dissecting some and stuffing others. He also spent a great deal of time surveying the scenery and found evidence to support the theory of geological uniformitarianism*—the idea that the earth has experienced a series of physical changes over time by means of natural processes, first theorized by Scottish geologist James Hutton* in 1785.

Darwin returned in 1836, becoming an instant celebrity after publishing his journal as a travel book, *Journal of the Voyage of the Beagle*. He earned the respect of his father and his peers, and was elected secretary of the Geological Society. During the same period, he struck up friendships with the biologist Thomas Henry Huxley* and the geologist Charles Lyell,* people who would be invaluable to his thought and career.[5]

In 1839, Darwin married his cousin Emma Wedgwood.* Less than a year later, he became ill and the couple moved to Down House in Kent. Perhaps this illness gave him the excuse he was looking for to withdraw from the distractions of society and concentrate on his scientific research.[6] Darwin had ten children, three of whom died in childhood; he was as devoted to his family as to the natural sciences. After the publication of *On the Origin of Species*, poor health almost invariably kept him away from the ensuing heated public debates, but he was kept up to date with events by letter correspondence with his close friends, such as Huxley. In Kent, he lived a quiet life, rarely socializing and instead focusing on family life and writing books and scientific papers.

Author's Background

In his youth, Darwin collected minerals, birds' eggs, and insects—pursuits regarded by his father, the scientist and medical doctor Robert Waring Darwin,* as a self-indulgent waste of time. There were several other scientists in the family, among them his grandfather Erasmus Darwin,* a physician and naturalist, and his cousin Francis Galton,* a statistician and anthropologist. His maternal great-grandfather was the famous entrepreneur Josiah Wedgwood,* a potter and prominent abolitionist*—an activist in the struggle to abolish slavery. Both families were predominantly Unitarians, a sect of Protestantism. Darwin's dislike of slavery and his support for the Great Reform Act of 1832* (legislation that granted some previously disenfranchised men the right to vote) reveal his general tendency toward liberal* and reformist views.

With personal financial security, and therefore more leisure time for exploration than the average Victorian, Darwin was able to pursue his interest in natural history. His social standing gave him access to some of the greatest scientific, philosophical, and literary minds of the day,[7] including the scientist and mathematician John Lubbock* and the social theorist Harriet Martineau. In this environment, his ideas blossomed, backed up by his practical experiments in botany, zoology (the study of animals), and biological inheritance* (the various characteristics that can be passed from parent to child). At the time of the text's publication in 1859, Darwin was given further professional and personal support by Charles Lyell, Thomas Huxley, and the botanist Joseph Dalton Hooker.* Despite the religious controversies surrounding his theory of evolution,[8] Darwin maintained an academic career of note to the end of his life.

NOTES

1 Edward J. Larson, *Evolution: The Remarkable History of a Scientific Theory* (New York: Modern Library, 2004).

2 Cynthia Delgado, "Finding the evolution in medicine," *National Institutes of Health Record*, 58.15 (2006), 1–8.

3 Cynthia Delgado, "Finding the evolution in medicine," 1–8.

4 Browne, *Charles Darwin: A Biography. Vol. 1: Voyaging* (Princeton: Princeton University Press, 1996), 75–6.

5 Browne, *Voyaging*, 355–6.

6 Janet Browne, *Charles Darwin: A Biography. Vol. 2: The Power of Place* (London: Pimlico, 2003), 5.

7 Browne, *The Power of Place*, 5.

8 Browne, *The Power of Place*, 5.

MODULE 2
ACADEMIC CONTEXT

KEY POINTS

- The natural sciences are concerned with understanding the physical world.

- During the intellectual movement known as the Enlightenment of the seventeenth and eighteenth centuries, in which society and scholarship took a turn toward rationality, several scientists had considered the development of new species but were unable to explain fully how it occurs.

- *On the Origin of Species* answers the then-pressing question of how species change gradually over time by proposing a mechanism called "natural selection."

The Work in its Context

It would be difficult to think of a book that embodies the Victorian era more fittingly than Charles Darwin's *On the Origin of Species by Means of Natural Selection* (1859). In the nineteenth century, England was bursting with ideas and social innovations that challenged both state and Church; among them were the Great Reform Act of 1832, which granted the vote to many thousands of men previously denied it, debates concerning the human being's place in nature, and the abolitionist* (anti-slavery) movement. The blooming of new social and scientific ideas, nonconformist* sects (religious faiths that confronted the Anglican Church), and the anti-alcohol temperance* movements combined to challenge the post-Enlightenment world and its dominant ideologies.

There had already been rumblings on this front, with European and American calls to give the vote to all men (universal male suffrage)

> 66 When we contemplate every complex structure
> and instinct as the summing up of many contrivances,
> each useful to the possessor, nearly in the same way
> as when we look at any great mechanical invention ...
> when we thus view each organic being, how far more
> interesting, I speak from experience, will the study of
> natural history become! 99
>
> Charles Darwin, *On the Origin of Species by Means of Natural Selection*

and the mid-seventeenth century emergence of liberalism* (the protection of an individual's liberty by a system of checks and balances on government). Ideas as subversive as those found in *On the Origin of Species*, which suggests that nature, not God, created life, challenged authority at the highest imaginable level. The work was perfectly placed to question the dominant paradigm* (the model by which knowledge was attained and understood): the idea that organisms are static and never change, and were created by a divinity who also gave human beings dominion over them. Many scientists had long wanted to challenge this model—but few had dared.

Overview of the Field

According to literalist* Christian belief, the earth was shaped by the damage inflicted on it when God punished man with a great flood; the earth is a static ruin and has not changed ever since. The species that exist today have always existed, and their extinction or change is inconceivable.

Christian doctrine was not the only important intellectual influence on the paradigm of an unchanging earth and static species. The philosophy of the ancient Greek thinker Plato* had a similar effect. According to Plato, the universe contains fixed ideal types of everything that can be seen and felt (or indeed imagined); these ideal types are hidden from us here in the everyday world, where only their distorted variations are visible. When this thinking, termed essentialism,* is applied to nature, biological change becomes impossible: although wild elephants, for example, might vary in height, contained hidden within each one lies a blueprint of the ideal elephant[1] and "change" is nothing but a distorted variation.

This view was called into question in the eighteenth century, when geologists began to discover certain types of shellfish that had existed in the past but were no longer alive—a challenge to both the biblical and the Platonic views of the world. Such scientists continued to find more extinct life forms. In order to reconcile these findings with religious orthodoxy, Christians felt they must be the result of intermittent catastrophes: there had not been just one flood but many, the last Great Flood* being Noah's.

By the end of the eighteenth century, evidence in favor of evolution was emerging in other disciplines too. The year before *On the Origin of Species* was published, an almost complete skeleton of the dinosaur *Hadrosaurus* was found in the US state of New Jersey. A decade earlier, in 1841, the British naturalist Richard Owen* had presented dinosaurs as a separate taxonomic* group (an order of animals unique to itself), coining the name *Dinosauria*.[2] Darwin, in fact, left dinosaurs out of his text, since only a few specimens had then been found; no one knew much about them or if they had left any living descendants. Stronger supporting evidence for evolution came from comparative anatomy*—the study of the comparison and contrast of the anatomies of different species.

Many vertebrate animals—animals with a spine—have a very similar anatomical structure. So the same five-digit structure is observed in the human hand as in the wing of a bat, which has five digits ("fingers") still present in its anatomy. Furthermore, comparative anatomists observed that there are phases in the development of the human embryo that look identical to those of birds, reptiles, and mammals. All this evidence was inconsistent with the notion of intelligent design* (that there is a divine "designer" of the earth's organisms), and Darwin would come to adopt it as supporting evidence for common descent.

Academic Influences

Theorizing evolution was no new matter– philosophers and thinkers ranging from the Greeks Anaximander of Miletus* (610–546 BCE) and Empedocles* (490–430 BCE) to medieval Arabic scholars such as al-Jahiz* (776–868) and Nasir al-Din al-Tusi* (1201–1274), right up to the French naturalist and mathematician George Louis Leclerc* (1707–1788), had accurately hypothesized many elements of modern evolutionary theory as we understand it today. For example, al-Tusi had proposed that minerals evolve into animals, and then humans, and those that are fittest survive, while Leclerc argued that that living things change through time and that humans and apes are related. It was not evolutionary thought that was new, *per se*; what Darwin and Wallace proposed in 1858 was a working mechanism that explained species change.

Indeed, many more recent scholars had influenced Darwin, and their ideas appear in *On the Origin of Species*; in particular, the Scottish geologists Robert Chambers* and Charles Lyell. Chambers' evolutionary ideas had been published anonymously in a work called *Vestiges of the Natural History of Creation*; Lyell further developed Hutton's theory of uniformitarianism*—the idea that the earth has undergone a series of physical changes by means of natural processes over geological time. Darwin had read Chambers's work in great detail (if critically), and, inspired by Lyell, thought that if the physical environment could change, then so could animals and plants. Indeed, they had to adapt to fit the changed environment, or become extinct.

The French biologist Jean-Baptiste Lamarck also was of key importance to Darwin. Devoted to the doctrine of the Great Chain of

Being*—an arrangement of natural types ordered from the simplest to the most complicated, also known by its Latin name *scala naturae*— Lamarck saw the idea as a staircase that species ascend, striving to become more complex. He theorized that individuals could modify their bodies in response to environmental problems and pass these new beneficial traits on to their offspring.[3]

Another major influence was the pioneering statistician Thomas Robert Malthus,* author of *An Essay on the Principle of Population*, first published in 1798. This text inspired Darwin (as well as Alfred Russel Wallace) to consider the possibility of a natural mechanism by which overpopulation is held in check.

Darwin's theory of natural selection could not have existed without the work of naturalists, anatomists* (specialists in the anatomy of organisms), and taxonomists* (specialists in the categorization of organisms), who helped him to catalogue and describe the vast hoard of specimens he had collected during his voyage on the HMS *Beagle*. The anatomist Richard Owen and the ornithologist John Gould* (an expert in birds) described some of these specimens. For example, it was Gould who realized that birds brought back from the Galapagos Islands that Darwin believed to be blackbirds, grosbeaks, and finches, were in fact 12 separate species of finches,[4] a revelation that influenced Darwin's conceptualization of species adapting to differing environments.

NOTES

1   Edward J. Larson, *Evolution: The Remarkable History of a Scientific Theory* (New York: Modern Library, 2004).

2   Deborah Cadbury, *Terrible Lizard: The First Dinosaur Hunters and the Birth of a New Science* (New York: Henry Holt, 2000).

3   Rebecca Stott, *Darwin's Ghosts: In Search of the First Evolutionists* (London: Bloomsbury, 2012), 194–7.

4   Adrian Desmond and James Moore, *Darwin* (London: Michael Joseph, 1991), 209.

MODULE 3
THE PROBLEM

KEY POINTS

- Organisms change and adapt in nature according to their specific environment and the organisms they are descended from.

- Biblical literalists (those who take the explanations offered by the Bible literally) disagreed; some esteemed scientists agreed with them.

- Darwin proposed that species do indeed change over time, and set out to discover the mechanism that underpins such changes.

Core Question

On the Origin of Species by Means of Natural Selection by Charles Darwin explains how all life on earth has evolved by the process of natural selection.

At the core of Darwin's investigation are the questions as to why some animals and plants have continued to exist for long periods of time while others have become extinct, and why there are similarities between many of these extinct forms and present-day species.

In the nineteenth century, it was assumed that the world and its inhabitants had always been the same and that everything was created by God. After his voyage around the world on the *Beagle*, Darwin began to question this doctrine, having seen firsthand evidence in the geology of South America and its fossils that contradicted this belief. By 1837, he was convinced that life had evolved, and he wanted to know how these evolutionary changes had taken place. Over time he would come to believe that new species come about not as the result of divine intervention but by adapting to changing conditions.[1]

> **❝** How have all those exquisite adaptations of one part of the organization to another part, and to the conditions of life, and of one distinct organic being to another being, been perfected? **❞**
>
> Charles Darwin, *On the Origin of Species by Means of Natural Selection*

The Participants

Christian doctrine claimed that the world had not changed since God created it, which placed a 6,000-year limit on the age of the planet. If the earth never changes physically, then there is no need for living things to alter either. In 1749, however, the French naturalist Georges-Louis Leclerc (later Comte de Buffon)* publicly questioned this, arguing that life had a history of its own and that the earth was more than 6,000 years old.[2] Buffon also observed similarities between humans and other primates such as chimpanzees, and suggested that human beings and the other apes share a common ancestry. While he raised the idea of biological change, Buffon did not, however, provide a coherent mechanism for how such changes occur.[3]

The naturalist Jean-Baptiste Lamarck presented another challenge to the biblical view in his *Philosophie Zoologique* (1809), offering a theory of evolution based on the mutation of species: his argument was that environmental challenges could force species to modify their bodies over time to gain personal advantage. He also suggested that offspring inherit these modifications from their parents[4], although he too got it wrong in the sense that he believed that modifications could be acquired in one's lifetime and then passed on to offspring. In 1826, the British naturalist Robert Grant began to speak publicly about evolution and promoted Lamarck's theory of transmutation* of species in Britain.[5]

In response to these challenges to creationism,* Bishop William Paley* argued that adaptations are the supernatural creation—indeed, proof—of God; the anatomist* Georges Cuvier* also criticized Lamarck's ideas, insisting that species are immutable (unchanging, and unchangeable). He did not reject evolution for religious reasons, however, but for reasons to do with evidence—the fossil record did not show any striving toward perfection.[6] For Cuvier, catastrophic changes were followed by acts of spontaneous generation, as with Aristotle's* hypothesis that logs can suddenly turn into crocodiles.

The Contemporary Debate

In the 1700s, geologists* began to discover rocks that contained within them records of previous extinction events (periods when large numbers of species die out simultaneously). Scientists at the time tried to accommodate Christian belief by introducing the theory of intermittent catastrophes, which posited that after each disaster, God had recreated the living forms on earth. The final catastrophe was Noah's great flood.* Such explanations nevertheless failed to explain why some animal types had perished during these events, while others had survived: a badger species existed in the Miocene Age, for example, that was almost identical to the badger of the present day.[7]

Building on the evidence from fossils found in ancient rocks, in 1788 the Scottish geologist James Hutton introduced the theory of uniformitarianism: the idea that the earth had undergone continuous physical change in the past and that the same transformative process is continuing to the present day. This change is constant and gradual rather than a series of catastrophic events. Hutton's theory was largely overlooked until the nineteenth century, when the geologist Charles Lyell sought to develop and popularize it.[8]

The French biologist Jean-Baptiste Lamarck made a forceful argument in his *Philosophie Zoologique* (1809). Having absorbed the ideas of essentialism from the thought of the Greek philosopher Plato,* according to which any animal or physical object possesses a set of attributes necessary to its identity and function, and also that of the Great Chain of Being, Lamarck believed that all living things are engaged in a constant struggle to reach ever-greater complexity. Their ultimate goal is to become as complex as man. As some organisms become more complex, gaps appear at the bottom of the ladder, to be filled by simple, spontaneously generated organisms. Lamarck thought that two forces direct this process: the inherent drive toward complexity, and the environment. Living things, he believed, rise to ecological challenges by modifying their bodies, and these changes are passed on to the next generation. Despite his misunderstanding that heritable modifications can be acquired in one's lifetime, Lamarck's argument was so persuasive that it remained the most influential evolutionary theory until that of Darwin and the Welsh naturalist Alfred Russel Wallace,* who came to very similar conclusions.[9]

However, Lamarck's theory of the transmutation of species was associated with the radical materialism* of the Enlightenment ("materialism" here means the assumption that all physical phenomena have physical causes), and was greeted with hostility. In his youth Darwin had read Bishop's Paley's *Natural Theology* (1809), a book written partly in response to this theory. In it, Paley declares adaptations to be the supernatural creations of God.[10] Indeed, the existence of such adaptations in nature provides one of the main philosophical arguments for the existence of God. This is known as the argument for "providential design."

Lyell was not an unequivocal supporter of the principles that fed Darwin's theory. Although he later supported Darwin professionally and personally, Lyell criticized Lamarck's theories in *Principles of Geology* (1830, 1833). Writing about uniform change in inorganic matter, Lyell refused to believe in the possibility of biological change. Instead, he proposed that each species has its "centre of creation" and is designed for a particular environment; species become extinct when the environment that supports them changes.[11] The Scottish writer Robert Chambers, on the other hand, *was* inspired by Lamarck and, in his *Vestiges of the Natural History of Creation* (1844) – a work which influenced both Darwin and Wallace to develop their theories – suggested that all complex life evolves from simpler forms.

NOTES

1 Edward J. Larson, *Evolution: The Remarkable History of a Scientific Theory* (New York: Modern Library, 2004).

2 Rebecca Stott, *Darwin's Ghost: In Search of the First Evolutionists* (London: Bloomsbury, 2012).

3 Georges-Louis Leclerc, Comte de Buffon, *Les Epoques de la Nature*, in *Histoire Naturelle, générale et particulière, avec la description du Cabinet du Roi* (Paris: imprimerie nationale, 1749–1788).

4 J. B. Lamarck, *Zoological Philosophy: An Exposition with Regard to the Natural History of Animals*, trans. Hugh Elliot (Chicago: University of Chicago Press, 1984).

5 Stott, *Darwin's Ghost*.

6 G. Cuvier, *Tableau elementaire de l'histoire naturelle des animaux* (Paris: Baudouin, 1798), available at https://archive.org/details/tableaulment00cuvi, 71. [Accessed February 16, 2016.]

7 Robert Chambers, *Vestiges of the Natural History of Creation* (London: John Churchill, 1844).

8 Larson, *Evolution*.

9 Larson, *Evolution*.

10 William Paley, *Natural Theology: or, Evidences of the Existence and Attributes of the Deity* (London: J. Faulder, 1809).

11 C. Lyell, *The Principles of Geology*, Vol 2 (London: Murray, 1830–3), chapter 2.

MODULE 4
THE AUTHOR'S CONTRIBUTION

KEY POINTS

- Scientists had long been looking for a working model to explain variation within and between species. It was Darwin (and his contemporary Alfred Russel Wallace) who hit on it at last.

- Natural selection provided a working mechanism to explain how different species come about and how they change over time.

- Darwin proposed that individual organisms vary in their characteristics, and that they compete with each other in a struggle for existence in a world of limited resources.

Author's Aims

In *On the Origin of Species by Means of Natural Selection,* Charles Darwin's aim was to explain adaptation and evolutionary change. Such a theory needed to account for the full diversity of life on earth and explain why organisms look and behave the way they do. The adaptations included physical changes in a part of the body, such as a wing, as well as a particular behavior—both being designed to improve an organism's chances of survival and reproduction. Such adaptations could not arise by chance, and they required an explanation.[1]

The concept of evolution has been around since the age of ancient Greece, and conceived of separately too in China and in the Middle Eastern caliphates. When Darwin came to publish *On the Origin of Species*, many philosophers and scientists had already accepted that animal and plant life evolves over time.[2] By the nineteenth century, the disciplines of geology,* embryology,* and anatomy* had

> **❝ Light will be thrown on the origin of man and his history. ❞**
>
> Charles Darwin, *On the Origin of Species by Means of Natural Selection*

accumulated plenty of evidence in favor of evolution. Scientists had attempted to explain how the various forms and varieties of animal and plant types had come about, but no one had managed to do so convincingly. Darwin changed this, offering a coherent argument understandable to scientists and nonscientists alike: organisms over-reproduce and then compete for resources and survival; those best adapted to a particular environment out-compete those less well adapted, and successfully pass on their useful adaptations to the next generation.

Approach

In *On the Origin of Species*, Darwin constructs a scientific argument using the comparative method—he presents the reader with a set of detailed observations and then uses reasoning to support his idea. In order to illustrate how varieties occur in nature, for instance, he uses examples from domestic animals. He explains how the selective breeding of dogs has resulted in various breeds that all look very different from one another, even though they are all descendants of a common ancestor. This shows that a handful of breeds can produce hundreds more—which, Darwin concludes, must also be possible in wild animals.

Darwin provides evidence, too, for his idea of gradualism.* Using eyes as an example, he explains how the eyes of mammals* might have started out as simple light-sensitive organs, as seen in invertebrates (animals without a spine).[3] Of course, as eyes are composed of soft tissue they do not appear in fossils, but Darwin argues his case in principle, inferring the stages of evolution that led to the human eye. In this way, he rejects the notion of intelligent design* and replaces supernatural intervention with natural selection.[4]

Contribution in Context

Darwin and his contemporary, the naturalist Alfred Russel Wallace, conceived an original theory for evolution that used evidence from the fields of geology, embryology, and anatomy. This was a quite a departure from the approach taken by Jean-Baptiste Lamarck and some other German scientists, who built on speculation and theorizing. It is also worth noting that both Darwin and Wallace started from human beings and worked back to animals when creating their theory of natural selection.

Although Darwin and Wallace are credited as co-discoverers of natural selection, a careful reading of the article they jointly published in 1858 shows they had different approaches to the theory. Only Darwin understood that competition is greatest within members of the same population, rather than between species. He also noted that species with a recent common ancestor tend to appear more similar, while those with a more distant common ancestor have fewer shared characteristics.[5] This led him to his principle of divergence, and how species evolve over time.

Wallace, on the other hand, emphasized the importance of the environment in shaping species.[6] He believed that food supply and predation are particularly influential on population growth, and concluded, "The numbers that die annually must be immense; and as the individual existence of each animal depends upon itself, those that die must be the weakest—the very young, the aged, and the diseased—while those that prolong their existence can only be the most perfect in health and vigor—those who are best able to obtain food regularly, and avoid their numerous enemies."

Contrary to Darwin, Wallace emphasized that an individual is deemed *weak* not on the basis of biological inheritance, but pure chance; the example he gives is the very old or young.

NOTES

1   Mark Ridley, *How to Read Darwin* (London: Granta Books, 2006).

2   Edward J. Larson, *Evolution: The Remarkable History of a Scientific Theory* (New York: Modern Library, 2004).

3   Ridley, *How to Read Darwin*.

4   Ridley, *How to Read Darwin*.

5   Michael Bulmer, "The theory of natural selection of Alfred Russel Wallace FRS," *Royal Society Journal of the History of Science* 59, no. 2 (2005): 125–36.

6   Bulmer, "The theory of natural selection," 125–36.

SECTION 2
IDEAS

MAIN IDEAS

KEY POINTS

- The main theme explored by Darwin is that individuals firstly vary within any population.

- This variation results in differing rates of survival and reproduction.

- Finally, individuals with traits most favorable to survival live, go on to reproduce, and pass on those traits to their offspring. Those species that do not survive to reproduce become extinct.

Key Themes

The core themes of Charles Darwin's *On the Origin of Species by Means of Natural Selection* are variations between organisms and within environments, and competition within populations for resources. Together, they form the mechanism of evolution, which Darwin termed "natural selection."

"Competition" in biology is an interaction between two or more organisms that require the same limited resources (such as food, water, or a mate, which the organisms require in order to grow, survive, and reproduce). As an organism cannot secure a resource if another organism has already consumed or defended it, competitors reduce the potential of others for growth, reproduction, or survival.

Exploring the Ideas

Darwin begins *On the Origin of Species* by illustrating that there is built-in variation within species, using everyday examples from among domestic animals and plants. He observes that some domestic animals

> **❝ We are not here concerned with hopes or fears, only with the truth as far as our reason allows us to discover it. ❞**
>
> Charles Darwin, *The Descent of Man, and Selection in Relation to Sex*

have an "extraordinary tendency to vary"[1]—notably the dog. Over several pages, Darwin describes the various dog breeds and their physical differences, speculating that they probably originated from just a handful of wild species. Darwin calls them varieties because it is possible to interbreed domestic dogs (*Canidae*), so they are classed as one species. From domestic animals he goes on to explore the natural variation that occurs in wild animals and plants. He admits that the domestic dog is a product of human manipulation—in the form of selective breeding, according to which a breeder seeks to develop traits such as size or intelligence by mating animals that present these characteristics—and would probably not have come about naturally. However, he stresses the fact that the variation observed in dog breeds must be inherent within its wild ancestors.[2]

Darwin's next theme is the "struggle for existence." Some individuals are better at competing for resources than others. Darwin reasons that the most successful will possess a variation of a particular trait or traits that gives them an advantage. These better-adapted individuals are likely to produce more offspring than others. Offspring inherit the favorable traits of their parents by means of biological inheritance. He could not describe the actual mechanism involved, because the concept of genes—the biological material that carries characteristics down the generations—had not yet been publicized by the Moravian monk Gregor Mendel*; but he argues nonetheless that this process of natural selection results in a population with more individuals better adapted to their environment at a particular period. It is important to note that Darwin uses the phrase "survival of the fittest"* as a metaphor; what he was trying to say is that not all individuals will contribute equally to the next generation. The word "fittest" is not a value judgment: in biology the word designates the individuals that survive and, importantly, have offspring.

Darwin uses selective breeding to introduce the theme of gradualism, showing how by small incremental steps, over successive generations, adaptations can arise. He admits that variations in domestic animals are not likely to be beneficial to the animals themselves, but are for "man's own use or fancy."[3] Here, the environment acting on the individual domestic animal is human culture, and these examples still illustrate that organisms have the capacity to change gradually over time—illustrating both the concept of gradualism and that individuals can inherit traits from their parents.

Language and Expression

On the Origin of Species was written for a well-educated general reader skilled in critical thought. It is not necessary to have a scientific background to understand the text, written as it is in the standard prose of the period.

Darwin presents his case in favor of evolution, implicitly rejecting creationism (the doctrine that the biblical account of the creation of the world and its animals is perfectly correct). He uses the term "descent with modification"* rather than "evolution." The latter came to be more widely used in the later part of the nineteenth century, appearing only in the sixth edition of 1872; the word "evolved" is only used once, in the work's very last sentence.[4]

Darwin uses the term "natural selection" to explain how organisms diversify and adapt. He makes only a few direct references to intelligent design, and writes solely about the nature of organisms in terms of their physical properties and nature. He does his best to avoid stating explicitly how his theory undermines the biblical account of the origins of life. His argument was clear to his readers, though, and

Darwin's contentious ideas rose from the 18th-century concept of positivism*—which argues that information derived from sensory experience, and interpreted through reason and logic, forms the only source of authoritative knowledge.

Throughout the book, Darwin uses analogies, such as his comparison of domestic and wild animals; but it was not only his references to domesticated animals, but to domestic life, that enabled readers to identify with his writings easily.[5] One wonderful example of this is when he describes how he uses a teacup to scoop up mud from a local pond. In this way his ideas are expressed informally, and were accessible to the Victorian reader. He aimed to reach a wide audience: "I sometimes think that general and popular Treatises are almost as important for the progress of science as original work."[6] Darwin succeeded: *On the Origin of Species* is now a world-famous work.

Indeed, in 2012, *On the Origin of Species* was cited by 90 per cent of the participants in a *New Scientist* poll as the most influential science book ever.[7] Its status as a frequently banned book[8] also gives testament to its lasting influence and sometimes-controversial centrality regarding what it means to be a human being.

NOTES

1   Charles Darwin, *On the Origin of Species by Means of Natural Selection, or the Preservation of Favored Races in the Struggle for Life*. Introduction and Notes by Gillian Beer (1996, 2008), (Oxford: Oxford University Press, 1860).

2   Darwin, *Origin*.

3   Darwin, *Origin*.

4   Mark Ridley, *How to Read Darwin* (London: Granta Books, 2006).

5   George Levine, *Darwin the Writer* (Oxford: Oxford University Press, 2011).

6   Charles Darwin, in a letter to Thomas Huxley (1865), in *Darwin's Island: The Galapagos in the Garden of England* by Steve Jones (London: Little Brown, 2009).

7   CultureLab. "Top 10 Most Influential Popular Science Books". *New Scientist,* October 1, 2012. Available at http://www.newscientist.com/ blogs/culturelab/2012/10/top-10-most-influential-popular-science-books. [Accessed July 7, 2013.]

8   Daily Lit "Banned Books" (popular), available at http://www.dailylit.com/tags/ banned-books/popular/1. [Accessed July 7, 2013.]

MODULE 6
SECONDARY IDEAS

KEY POINTS

- Darwin's secondary ideas are sexual selection (the process by which evolution occurs through the selection of a mate), gradualism of species, and speciation (the formation of distinct species).

- These secondary ideas, especially sexual selection and speciation, are still relevant to contemporary evolutionary studies.

- The most important secondary idea is sexual selection, which is still an important factor in natural selection as understood in modern evolutionary theory.

Other Ideas

A key subordinate idea in Charles Darwin's *On the Origin of Species by Means of Natural Selection* is the component of natural selection known as "sexual selection." Here it is the environment of sexual competition and sexual choice that limits or favors particular traits. This theory partly explains a paradox: how is it that some characteristics evolve when they reduce an individual's chances of survival—the peacock's tail, for example?

Another important secondary idea concerns the origin of new species—speciation.* Darwin's concept of evolution by natural selection requires "slow and gradual accumulation of numerous, slight yet profitable" modifications—termed gradualism.[1] In chapter 6 he highlights a problem, "Firstly, why, if species have descended from other ancestral species by insensibly fine gradations [over a long period of time] do we not everywhere see innumerable transitional forms?"[2]

> ❝ The sexual struggle is of two kinds: in the one it
> is between the individuals of the same sex, generally
> the males, in order to drive away or kill their rivals,
> the females remaining passive; while in the other, the
> struggle is likewise between the individuals of the same
> sex, in order to excite or charm those of the opposite
> sex, generally the females, which no longer remain
> passive, but select the more agreeable partners. ❞
>
> Charles Darwin, *The Descent of Man, and Selection in Relation to Sex*

Addressing here the then-significant problem of the scarcity of
transitional types in the fossil record, he goes on to discuss at great
length how there must be some method in natural selection that leads
to the generation of distinct species.

Exploring the Ideas

There are two forms of sexual selection. In intrasexual selection,
members of the same sex compete for opportunities to mate; red deer
stags, for example, compete for access to females by clashing antlers. In
intersexual selection,* members of the opposite sex are attracted to
particular characteristics, such as extravagant and colorful tail feathers,
as seen in male peacocks. If the peacock males chosen by the peacock
hens are genetically different from those of their rivals, then natural
selection occurs.[3]

According to Darwin, evolution by natural selection is a gradual
process. New species are formed by slow incremental changes that
take place over long periods of time. One of the major difficulties for
Darwin was the absence in the fossil record of evidence of these
changes, in the form of intermediary types.

Until recently, it was accepted that scientists were unable to find a
fossil record for many intermediate forms. The evolutionary biologists

Stephen Jay Gould and Niles Eldredge* proposed an explanation for these gaps, suggesting that they are real, and represent periods of stability when species did not change much for millions of years. They believed these periods of stability are then followed by times of rapid change that result in new species, termed "punctuated equilibrium."*[4]

According to this theory, changes resulting in a new species do not normally come from small, gradual shifts in the mainstream population, but occur instead in a small subset of the population, such as those living at the edge of the habitat or in a small isolated group. When the environmental conditions change, these "peripheral" or "geographic isolates" experience intense selection and speedy change due to both the altered environment and their small population size. They do not leave fossils reflecting the intermediate stages because their numbers are relatively few, and because of their isolated location. These new successful types can then spread out across the geographic area of the ancestral species.[5] The theory of punctuated equilibrium does not imply that evolution *only* happens in rapid bursts. Moreover, in the last decade, many more transitional fossils have indeed been found, including within our own species.[6] Many paleontologists* now believe it is not an either/or situation: the mechanisms of both gradualism and punctuated equilibrium working at different points in time.[7]

Does speciation imply that organisms exist as definite types? Darwin had reservations about the existence of distinct species: "Nor shall I here discuss the various definitions which have been given of the term species. No one definition has as yet satisfied all naturalists; yet every naturalist knows vaguely what he means when he speaks of a species."[8] In the second half of the twentieth century, over 20 different definitions of "species" were proposed.[9] Some argue that such a large number of definitions is itself proof that distinct species do not exist, and prefer to view living organisms as part of a gradually changing continuum rather than as distinct entities. Categories such as species, they say, are more a product of the human mind, which has a tendency to classify things for simplicity's sake, rather than being a true reflection of nature.[10]

Overlooked

Despite its importance to evolutionary theory, the concept of sexual selection was largely ignored for nearly a century, because people found it difficult to comprehend how animals could exhibit choice. Darwin himself had a problem with this notion, as animals were then regarded as robots at the mercy of their instincts, though he noted later that earthworms can make choices.[11] So, though scientists nowadays understand that choice as a mechanism can exist,[12] it was a difficult conceptual jump to make in Darwin's time.

Sexual selection has become central to modern evolutionary biology and behavioral ecology* (a field in which animals' environments are used to explain their behavior). This is in part due to the British statistician Ronald Fisher,* who developed a model for sexual selection in 1915. He suggested that traits such as a male peacock's tail can evolve if there is a genetic basis for both the trait itself (the tail) and the sexual preference for the exaggerated version of that trait. If females carry a gene for the preference and their sons have a gene enabling them to develop the preferred trait, this results in a proportional increase in both the trait and the preference for it.[13] Fisher's work, however, was overlooked until the 1970s when it was discovered that female choice does indeed exert a very powerful effect on male traits.[14]

NOTES

1   Charles Darwin, *On the Origin of Species by Means of Natural Selection, or the Preservation of Favoured Races in the Struggle for Life* (London: John Murray, 1859), 103.

2   Darwin, *Origin*.

3   J. R. Krebs and A. Kacelnik, "Decision-making," in *Behavioural Ecology: An Evolutionary Approach* (Oxford: Blackwell Scientific, 1991), 105–36.

4   Niles Eldredge and S. J. Gould. "Punctuated equilibria: an alternative to phyletic gradualism", in T.J.M. Schopf, ed., *Models in Paleobiology*, ed. T. J. M. Schopf (San Francisco: Freeman Cooper, 1972), 82–115.

5   Gould, "Punctuated equilibria," 82–115.

6   W.A. Haviland and G.W. Crawford, *Human Evolution and Prehistory* (Cambridge, MA: Harvard University Press, 2002).

7   Prothero, Donald R. *(1 March 2008)*. "Evolution: What missing link?" *New Scientist*. London: Reed Business Information. 197 (2645): 35–41. ISSN 0262-4079. doi:10.1016/s0262-4079(08)60548-5. Retrieved May 13, 2015.

8   Darwin, *Origin*.

9   J. A. Mallet, "Species definition for the modern synthesis," *Trends in Ecology and Evolution* 10 (1995): 294–9.

10 Daniel Elstein, "Species as a Social Construction: Is Species Morally Relevant?" *Journal for Critical Animal Studies* 1, no. 1 (2003): 53–71. See also John Wilkins, *Species: A History of the Idea* (Oakland: University of California Press, 2011).

11 Darwin, *The Formation of Vegetable Mould, through the Actions of Worms, with Observations on Their Habits*, John Murray, London, UK, 1881.

12 Krebs, "Decision-making," 105–36.

13 Malte B. Andersson, *Sexual Selection*: *Monographs in Behavior and Ecology* (Princeton : Princeton University Press, 1994).

14 Andersson, *Sexual Selection*.

MODULE 7
ACHIEVEMENT

KEY POINTS

- Darwin and Wallace were the first to succeed in proposing a mechanism for explaining species change, which Darwin expanded upon. Natural selection remains the overriding paradigm (conceptual model) today.

- His work of over 21 years allowed him to present the strongest arguments for natural selection. *On the Origin of Species* went on to influence many other disciplines, including the life sciences.

- Darwin's ignorance of the mechanisms of inheritance through the biological material now known as genes—a field of inquiry that was not established until 1906—meant that he misunderstood how traits are passed down the generations.

Assessing the Argument

Charles Darwin's *On the Origin of Species by Means of Natural Selection* argues one key point, that natural selection is the driving force behind evolution: organisms over-reproduce and then compete for resources and survival, and then those best adapted to a particular environment out-compete those less well adapted, and successfully pass on their useful adaptations to the next generation. Darwin made his case by considering shared ancestry,[1] pointing to the fact that animals living in close proximity to each other tend to share a closer ancestry.[2] He also introduced the concept of gradualism as it applies to organisms.[3]

Knowing that he was courting controversy with the text's publication, Darwin devoted one chapter to answering all the questions he thought would arise, naming it "Difficulties on Theory."

> ❝ It will live as long as the 'Principia' of Newton ... Darwin has given the world a new science and his name should in my opinion stand above that of every philosopher. ❞
>
> Alfred Russel Wallace, Wallace Letters Online

While Darwin worked on his unpublished theory for twenty-one years until the time of the work's full publication in 1859, Charles Lyell, Thomas Henry Huxley,* and the botanist* Joseph Hooker gave Darwin their professional and personal support. These were invaluable friends, who helped to establish Darwin's intellectual priority*—the acknowledgment that it was Darwin's work, and not anyone else's, that had made the greatest step forward, thereby making sure he received the credit for his major discovery.

This support became critical in 1858 when it emerged that the naturalist Alfred Russel Wallace had also hit upon the same theory— natural selection—as the mechanism for species change. It was feared that Darwin might lose his intellectual priority to the younger man. Later, Lyell and Huxley would become a defense-and-attack team when *On the Origin of Species* became suddenly—and internationally— controversial. Darwin maintained his notable academic reputation and went on to publish three more major works on evolutionary theory.

Achievement in Context

Darwin was certainly not the first person to raise the topic of evolution; it had been a popular topic ever since the Enlightenment.[4] His own grandfather Erasmus Darwin published several works on the matter, including his famous poem "The Origin of Society" (1803), on natural history and the relatedness of all life forms. Another celebrated precursor was the French biologist Jean-Baptiste Lamarck, who in 1809 proposed that species could transform and were not, then, fixed entities. The power of Lamarck's writing was such that he made belief in evolution a respectable position to hold. Naturalists Étienne Geoffroy Saint-Hilaire* and Robert Grant went on to become advocates of this idea, as did the Scottish writer Robert Chambers who, in his anonymously published *Vestiges of the Natural History of Creation* (1844), suggested that all complex life evolves from simpler forms. Naturally, the clergy opposed *Vestiges* because it called traditional religious views into question, while scholars such as Adam Sedgwick criticized it for its superficiality on scientific grounds.[5] Nonetheless, *Vestiges* remained extremely popular among the general public, partly because of its accessible style,[6] and it undoubtedly prepared Victorian society for the idea of evolution.[7]

The first print run of *On the Origin of Species* was of 4,200 copies, which sold out on its first day; five further editions were published during Darwin's lifetime. While Chambers' *Vestiges* continued to outsell it,[8] *On the Origin of Species* ultimately had the most impact and was considered more respectable, partly due to the efforts of Darwin's close friends, and partly because of Darwin's own position as a notable scientist.[9]

Limitations

Darwin's ideas have at times been applied to social policy. Darwin himself had already applied his theory of natural selection more broadly. In *The Descent of Man, and Selection in Relation to Sex* (1871) he wrote, "All ought to refrain from marriage who cannot avoid abject poverty for their children," referring to the struggle for existence. "Otherwise he would sink into indolence, and the more gifted men would not be more successful in the battle of life than the less gifted."[10]

Such statements—some taken much further by eugenicists and political fascists such as the Nazis*—gave rise to social Darwinism*: the theory that, as with other species, humans are subject to natural selection, with some races being more genetically advanced than others. In this approach, biological concepts, like the survival of the fittest* (Darwin used the afore-noted phrase "struggle for existence"), are applied more broadly to economics, politics, and sociology. The idea that societies start out as "primitive" and advance toward civilization was first proposed by the English biologist Herbert Spencer,* who also developed the notion that certain (Western) cultures were superior.

Eugenics*—a scientific method aimed at improving the qualities of humans through selective breeding—developed from this idea. It was Darwin's cousin, Francis Galton,* who first proposed the science of eugenics in 1883. He advocated the imposition of social controls including sterilization for those deemed "unfit." The idea that "unfit" individuals, such as those afflicted with physical or mental disabilities, should be stopped from reproducing was implemented in several countries, including Britain, Canada, Germany, Sweden, and the United States, and the policies remained in place until the 1970s and 1980s, in some cases.[11]

The German philosopher Friedrich Nietzsche* warned that, taken out of context, the ideas in *On the Origin of Species* could be interpreted as nihilistic—that is, suggesting that our existence is meaningless. This could absolve the individual of all moral or social responsibility, for if life has no purpose, why value it? This position is a misunderstanding of natural selection, for which life indeed has a purpose, which is to survive and reproduce. By agreeing that if something has evolved, it must be good, we commit what British philosopher G.E. Moore* argued is a "naturalistic fallacy,"*[12] because good has been defined as something other than itself, rather as *the survival of the fittest.*"[13]

Another limitation rose from the fact that Darwin was not aware of Gregor Mendel's work and did not understand genetic principles. Only once Mendel's work was "rediscovered" in 1900 (after Darwin's death) did it become clear exactly how complementary Mendel's Laws of Inheritance and Darwin's theory of natural selection were to each other.

In biology, there are two kinds of variation. The first includes any difference between cells, individual organisms, or groups of organisms as a result of genetic* differences (differences in genes,* which decide an organism's properties). This difference in a genotype* is known as "genotypic variation." The second takes into account the influence of the environment on genetic potential; this difference in a phenotype* is known as "phenotypic variation." Variations can be observed either in physical appearance, such as height, or in behavior. "Biological inheritance" describes the process by which genes are passed on from parent (or ancestor) to offspring.

NOTES

1   Charles Darwin, *On the Origin of Species by Means of Natural Selection, or the Preservation of Favored Races in the Struggle for Life.* Introduction and Notes by Gillian Beer (1996, 2008), (Oxford: Oxford University Press, 1860).

2   Darwin, *Origin.*

3   Mark Ridley, *How to Read Darwin* (London: Granta Books, 2006).

4   Edward J. Larson, *Evolution: The Remarkable History of a Scientific Theory* (New York: Modern Library, 2004).

5   Adam Sedgwick, "Review of *Vestiges,*" *Edinburgh Review* 82 (July 1845): 1–85.

6   James A. Secord, *Victorian Sensation: The Extraordinary Publication, Reception, and Secret Authorship of Vestiges of the Natural History of Creation* (Chicago: University of Chicago Press, 2000).

7   Lois N. Magner, *A History of the Life Sciences* (New York; Basel: Marcel Dekker, 1994), 257–316.

8   Magner, *A History of the Life Sciences.*

9   George Levine, *Darwin the Writer* (Oxford: Oxford University Press, 2011).

10   Charles Darwin, *The Descent of Man, and Selection in Relation to Sex* (London: John Murray, 1871).

11   Dennis Sewell, *The Political Gene: How Darwin's Ideas Changed Politics* (London: Picador, 2009).

12   G. E. Moore, *Principia Ethica* (Cambridge: Cambridge University Press, 1993).

13   Julia Tanner, "The Naturalistic Fallacy," *Richmond Journal of Philosophy* 13 (2006): 1–6.

MODULE 8
PLACE IN THE AUTHOR'S WORK

KEY POINTS

- Darwin's life's work was to explain the natural world through an understanding of evolution.*

- *On the Origin of Species by Means of Natural Selection* was the foundation on which all Darwin's later theories were built.

- *On the Origin of Species and Descent of Man* together made Darwin's reputation as the founder of evolutionary theory; his name has become synonymous with the principle of evolution itself.

Positioning

From an early age, Charles Darwin had an inquisitive nature and found the natural world around him fascinating. It is clear from reading *On the Origin of Species* that he found the biblical explanation for the creation of this natural world unsatisfactory. In his early twenties, he read widely on various subjects, many of which argued in favor of the active role of the divine in the affairs of humankind; natural theology*—the idea that the beauty and complexity of nature is itself evidence for the existence of God—was the dominant scientific position at that time.

Darwin regarded these explanations as unreasoned; throughout his life, his purpose was to explain the natural world he observed around him. The theories presented in *On the Origin of Species* and his subsequent books were thoughtful and detailed accounts of his ideas.

The central questions found in Darwin's books took shape early on in his life, perhaps soon after his return from the voyage aboard

> **❝** As long as man and all other animals are viewed
> as independent creations, an effectual stop is put to
> our natural desire to investigate as far as possible the
> causes of Expression ... The community of certain
> expressions in distinct though allied species ... is
> rendered somewhat more intelligible, if we believe
> in their descent from a common progenitor. He who
> admits on general grounds that the structure and
> habits of all animals have been gradually evolved, will
> look at the whole subject of Expression in a new and
> interesting light. **❞**
>
> Charles Darwin, *The Expression of the Emotions in Man and Animals*

HMS *Beagle*.[1] In 1838, he read the Scottish anatomist and philosopher Charles Bell's* *Essays On The Anatomy And Philosophy Of Expression* (1824), which concluded that man's emotions are unique.[2] From the notes in the margins of Darwin's original copy of Bell's book, it is clear that he did not agree,[3] and over the course of 30 years he gathered evidence that would eventually appear in *The Expression of the Emotions in Man and Animals* (1872). In this book Darwin uses descent from a common ancestor (evolution via gradualism) to explain similarities between the emotions of humans and other animals.

Integration
Evolution is the overarching theme in all Darwin's major texts: *On the Origin of Species* (1859), *The Variation of Animals and Plants Under Domestication* (1868), *The Descent of Man* (1871), and *The Expression of the Emotions in Man and Animals* (1872). In each book, Darwin explains how natural selection has shaped humans and other animals we see today.

In the two-volume *Variation of Animals and Plants*, Darwin expands his discussion about inheritance and variation in plants and animals. As noted previously, Darwin was not aware of Gregor Mendel's obscure work on biological inheritance. Instead, he introduces his theory of pangenesis*—the idea that small pieces of information in the form of gemmules (hypothetical particles) float from all parts of the adult body and go on to make the information that forms the offspring. In the very last paragraph of this book he argues more directly than in *On the Origin of Species* against the belief in evolution by design: "An omniscient Creator must have foreseen every consequence which results from the laws imposed by Him … [C]an it with any greater probability be maintained that He specially ordained for the sake of the breeder each of the innumerable variations in our domestic animals and plants;—many of these variations being no service to man, and not beneficial, far more often injurious, to the critters themselves? Did He ordain that the crop and tail-feathers of the pigeon should vary in order that the fancier might make his grotesque pouter and fantail breeds? … However much we may wish it, we can hardly follow … that variation has been led along certain beneficial lines."[4]

The Descent of Man expands on the ideas in *On the Origin of Species*, introducing further evidence for evolution in plants and animals. Darwin also sets out new details about human origins, focusing on the evolution of human mental and moral faculties. He argues that the difference between the mental faculties of humans and those of other animals is one of degree rather than kind. He also expands on his theory of sexual selection*—a form of natural selection in which the selection of a mating partner plays the principal role—that had previously figured in *On the Origin of Species*.[5] *The Descent of Man* was considered equally thought-provoking and scandalous.

Further evidence for human evolution from an ape-like ancestor appears in Darwin's subsequent book *The Expression of the Emotions in*

Man and Animals: "With mankind some expressions, such as the bristling of the hair under the influence of extreme terror, or the uncovering of the teeth under that of furious rage, can hardly be understood, except on the belief that man once existed in a much lower animal-like condition."[6]

Clearly, Darwin spent much of his life arguing in favor of evolution as a framework for explaining adaptations in humans and other animals. His books, and the influential way in which he considered human beings to be animals, revolutionized thinking in both the sciences and the humanities.

Significance

The debate about evolution in the 1850s was heated. When Darwin published in 1859, he transformed this debate largely because he was able to take all the current information and forge it into a coherent argument.

Darwin's first converts were his close friends Thomas Henry Huxley and Joseph Dalton Hooker, who then became vocal advocates for evolution. Both men were academic free thinkers who succeeded in their campaign to remove the influence of religious dogma from science. They backed the liberal* Anglican movement, which accepted evolution and opposed the traditionalists who had rejected it.[7] That said, Darwin also received full intellectual support from highly religious friends such as the American botanist Asa Gray.*

Darwin's influence soon stretched beyond academic circles. When the role of archbishop of Canterbury became vacant in 1896, it was filled by Frederick Temple,* an affirmed supporter of evolution.[8] When Darwin died, he was buried in London's Westminster Abbey near the great scientist Isaac Newton,* a mark of great public prestige, and both social and religious acceptance.[9]

Today the theory of natural selection is the unifying concept of the life sciences.

NOTES

1   Mark Ridley, *How to Read Darwin* (London: Granta Books, 2006), 8–15.

2   Charles Bell, *Essays on the Anatomy and Philosophy of Expression* (Montana: Kessinger Publishing, 2008).

3   Ridley, *How to Read Darwin*.

4   Charles Darwin, *The Variation of Animals and Plants Under Domestication* (London: John Murray, 1868).

5   Charles Darwin, *The Descent of Man, and Selection in Relation to Sex* (London: John Murray, 1871).

6   Charles Darwin, *The Expression of the Emotions in Man and Animals* (London: John Murray, 1872).

7   Edward J. Larson, *Evolution: The Remarkable History of a Scientific Theory* (New York: Modern Library, 2004), 79–111.

8   Philip Kitcher, *Living with Darwin: Evolution, Design, and the Future of Faith* (New York; Oxford: Oxford University Press, 2007).

9   Janet Browne, *Charles Darwin: A Biography. Vol. 2: The Power of Place* (London: Pimlico, 2003), 5.

SECTION 3
IMPACT

MODULE 9
THE FIRST RESPONSES

KEY POINTS

- On publication, *On the Origin of Species* was criticized because species change opposed the prevailing belief that organisms are fixed, and because it did not mention a divine Creator.

- The scientific mainstream accepted that species change exists; Darwin modified subsequent editions to include references to a Creator.

- The anatomist Thomas Huxley* undertook a spirited public defense of *On the Origin of Species*, as did religious friends such as the clergyman Charles Kingsley* and the botanist Asa Grey.

Criticism

Charles Darwin's *On the Origin of Species by Means of Natural Selection* and his theory of evolution provoked widespread criticism, principally focusing on his omission of an intelligent Creator—God. The concept of changing species was considered blasphemous, too, as it implied that God had created beings that were not perfect. It was not just the Church that disapproved. Darwin's close friend and mentor, the geologist* Adam Sedgwick, was disappointed in his pupil and feared the moral implications of the theory.[1]

Professional envy also played a part. In an attempt to discredit the theory, the anatomist* Richard Owen anonymously penned a long and malicious article in the *Edinburgh Review* in April 1860. Owen stated that Darwin's proposed mechanism of natural selection* was wrong, arguing for "the continuous operation of the ordained becoming of living things."[2]

> ❝ Science demonstrates incessant past changes, and dimly points to yet earlier links in a more vast series of development of material existence; but the idea of a beginning, or of creation, in the sense of the original operation of the divine volition to constitute nature and matter, is beyond the province of physical philosophy. ❞
>
> Rev. Baden Powell, *Philosophy of Creation*

There was also more general criticism on scientific grounds, some of which Darwin took seriously enough to address and refute in later editions. The first came from the zoologist* St. George Jackson Mivart, who argued that natural selection did not explain the early stages in the development of an organ such as the eye.[3] If the eye came about through small incremental steps, he argued, then there must have been a stage when it had no useful function and so gave no selective advantage.

Another objection came from the Scottish engineer Fleeming Jenkin in 1867. He suggested that in a large population, an adaptive trait existing among a few individuals would soon disappear when they interbred with others who lacked the feature. Jenkin supposed that genetic factors would divide during interbreeding and that a favorable trait would be diluted in subsequent generations.[4]

Responses

Darwin addressed the theological criticism in the second edition of *On the Origin of Species*, published just two months after the first in 1860, adding a few sentences on the Creator. He also quoted support for his idea of natural selection by a "celebrated cleric"— Reverend Charles Kingsley—whom he did not name but who, he wrote, "has gradually learnt to see that it is just as noble a conception of the Deity to believe that He created a few original forms capable of self-development into other and needful forms, as to believe that He required a fresh act of creation to supply the voids caused by the action of His laws."[5]

In response to Mivart's critique, Darwin argued that an organ could be beneficial during its early stages of development. For example, the eye would have started out as a light-sensitive eyespot. Over time, other developments would have occurred, granting greater benefit, and so through small steps, the eye evolved to become the complex organ it is today. Now, we understand that Darwin's explanation is correct: "a fortuitous novelty may confer a subtle advantage." During his lifetime, though, this issue remained a problem and one of the factors that meant natural selection was doubted by some as a theory to explain evolution.[6]

Darwin was also unable to show evidence for a mechanism of inheritance. In fact, less than a year after Fleeming Jenkin raised this question, Gregor Mendel, the founder of genetics, published a paper that would have answered his objections. Unfortunately, Mendel's work did not gain acceptance until its rediscovery in 1900. In his 1866 article, Mendel showed the action of "invisible factors" (or genes), in coding visible traits. He showed that these "factors" were indivisible and did not, therefore, blend during interbreeding as Jenkin had presumed.[7]

Following the objections, Darwin decided it was necessary to add some supplementary processes to the theory of natural selection in order to show that evolution could take place at a faster pace. In the sixth and final edition of *On the Origin of Species*, he writes that natural selection "aided in an important manner, by inherited effects of the use and disuse of parts; and in an unimportant manner, that is in relation to adaptive structures, whether past or present, by the direct action of external conditions."[8]

Here, Darwin reverts to the inheritance of acquired characteristics as first proposed by the French biologist Jean-Baptiste Lamarck. He also proposes a new theory, "pangenesis,"* according to which the sex cells (the sperm and ovum) absorb a set of particles, which represent all the organs and tissues of the adult body. In this way, characteristics acquired during the adult stage of life are passed on to offspring. This incorrect theory is similar to the one set out by the ancient Greek philosopher Democritus.*[9]

Conflict and Consensus

While *On the Origin of Species* was successful in convincing the scientific community to accept evolution, Darwin was not able to satisfy everyone that natural selection was the primary mechanism for species change, largely thanks to Jenkin's criticism. So after 1870, in the wider public natural selection somewhat fell out of favor and Lamarck's theory of transmutation—the idea that if an organism adapts its body to its environment during the course of its life, those changes are passed on to its offspring—became temporarily popular again.[10]

Other theories also sprang up at this time, such as orthogenesis,* now an obsolete biological hypothesis. Popularized by Theodor Eimer's* *Organic Evolution as the Result of the Inheritance of Acquired Characteristics According to the Laws of Organic Growth* (1890), orthogenesis suggests that organisms have an inborn tendency to evolve in a consistent and steady way, because of some internal mechanism or driving force.[11]

NOTES

1   R. Weikart, *From Darwin to Hitler: Evolutionary Ethics, Eugenics and Racism in Germany* (London: Palgrave Macmillan, 2004).

2   Mark Ridley, *How to Read Darwin* (London: Granta Books, 2006), 8–15.

3   St. George Jackson Mivart, *On the Genesis of Species* (Cambridge: Cambridge University Press, 2009).

4   Fleeming Jenkin, "Review of *The Origin of Species*," *The North British Review* 92, no. 46 (June 1867): 277–318.

5   Charles Darwin, *On the Origin of Species by Means of Natural Selection, or the Preservation of Favoured Races in the Struggle for Life,* 2nd edition (London: John Murray, 1860), 481.

6   Mario Livio, *Brilliant Blunders: From Darwin to Einstein: Colossal Mistakes by Great Scientists That Changed Our Understanding of Life and the Universe* (New York: Simon & Schuster, 2013).

7   Gregor Mendel, "Versuche über Pflanzenhybriden," *Verhandlungen des naturforschenden Vereins Brünn* (1866).

8   Charles Darwin, *On the Origin of Species by Means of Natural Selection, or the Preservation of Favoured Races in the Struggle for Life*, 6th edition (London: John Murray, 1872).

9   Edward J. Larson, *Evolution: The Remarkable History of a Scientific Theory* (New York: Modern Library, 2004).

10 Larson, *Evolution*.

11 Theodor Eimer, *Organic Evolution as the Result of the Inheritance of Acquired Characteristics According to the Laws of Organic Growth* (London: Macmillan, 1890).

MODULE 10
THE EVOLVING DEBATE

KEY POINTS

- *On the Origin of Species* has been deeply influential, notably in our understanding of human behavior—is it hardwired or socially constructed?—and the concept of "human uniqueness."*

- In Darwinian-influenced biological anthropology, humans are put in an animal context with other closely related species, and our behavior and variation are studied as they are in other animals.

- Darwin's work finally permitted scholars to make the necessary link between humans and our fellow primates; humans are now classified as animals and grouped in the family *Hominidae*, or the great apes*.

Uses and Problems

On the Origin of Species by Means of Natural Selection by Charles Darwin proposed that species evolve by natural selection. Fervent supporters of the theory – the co-discoverer Alfred Russel Wallace himself and the German naturalist August Weismann* – then founded a movement known as neo-Darwinism.*

Weismann scrutinized the concept that parents are able to pass on particular qualities (such as ability on the piano or a brawny body) to their offspring, going so far as to conduct experiments involving the cutting-off of mouse tails to see if scars on a parent would be passed on to the child. In a paper written in 1883, Weismann declared that acquired characteristics could not, in fact, be inherited. Sex cells (the collective name for sperm and egg) are segregated at an early stage in their development, so cannot be affected by any changes taking place

> **❝** If then, said I, the question is put to me would I rather have a miserable ape for a grandfather or a man highly endowed by nature and possessing great means and influence and yet who employs those faculties for the mere purpose of introducing ridicule into a grave scientific discussion—I unhesitatingly affirm my preference for the ape. **❞**
>
> Thomas Henry Huxley, "Letter to Dr. Dyster," September 9, 1860

in the parent's body once that has occurred.[1] By 1885, he had identified the nucleus of germ cells as the carriers of genetic information. Germ cells are biological cells that give rise to the gametes—the cells that develop from the fusion of sperm and egg—of any organism that reproduces sexually.

In 1889, in defense of natural selection, Wallace wrote *Darwinism*, a book that presented his own ideas on speciation (the process by which species are formed) and highlighted the importance of environmental pressures in forcing species to adapt to local habitats. When populations are isolated geographically and therefore have a smaller choice of mates, they begin to diverge until there are two separate species.[2] This theory came to be known as "the Wallace effect."[3] Current research supports this idea.[4,5]

In the early 1900s, the geneticist Nettie Stevens*, discoverer of the X-Y sex chromosome system, along with others such as the botanists* Hugo de Vries* and Carl Correns,* rediscovered the work of Gregor Mendel, the founder of genetics. Mendel's work showed that offspring retain distinct characteristics from each parent rather then a mixture of traits blended together. However, misunderstandings about Mendel's ideas led scientists to think that new characteristics or even species would suddenly appear.[6] De Vries went on to suggest that new species arise by a process of mutation*—abrupt changes in inheritable characteristics—and not by natural selection. In contrast to Darwin's idea of gradual change, de Vries thought that species evolve in sudden, dramatic changes. He based this "theory of mutation" on his work with the evening primrose plant. He saw that the original plant sometimes had offspring with significant visible differences, in, for example, leaf shape or plant height. De Vries then designated these as new species.

Essentially, de Vries had the right idea but for the wrong reasons; most of the variants he observed were due to aberrant chromosomal segregations (when paired chromosomes split and migrate to opposite ends of the nucleus), and were not mutations at all.

For several decades (1900–30), there were two opposing schools of thought. One upheld the Darwinian view of evolution by natural selection; the other believed that evolution was the result of a series of drastic mutations, as de Vries had proposed. In an attempt to confirm the latter theory, the US anthropologist Thomas Hunt Morgan* began working on the common fruit fly, and managed to establish a link between Mendel's work on the common pea plant and Walter Sutton's* work that identified genes as the carriers of hereditary information. Morgan's experiments showed that mutations did not suddenly create new species, but increased variations within a population.[7] The details of his work are published in *The Mechanism of Mendelian Heredity* (1915).[8]

From 1930 to 1950, scientists began to draw together the ideas of Mendel and Darwin. Darwin's theory of natural selection began to be more widely accepted, supported by new research from the fields of genetics and population statistics. This became known as the new synthesis period and gave rise to what is known as the modern evolutionary synthesis* and the field of evolutionary biology*— the inquiry into the mechanisms of inheritance conducted in the light of evolutionary theory.

Schools of Thought

Two major schools of thought emerged during the new synthesis period: modern evolutionary theory and sociobiology* (the field of science that suggests that human behavior has resulted from evolution). The Ukrainian evolutionary biologist Theodosius Dobzhansky* was a prominent figure in evolutionary theory. His *Genetics and the Origin of Species* (1937) combined Darwin's theory of natural selection together with Mendel's genetics and research from biological disciplines.[9]

In sociobiology, social behavior is explained by the notion that it has evolved to produce a beneficial outcome for the individual. The English scientist W. D. Hamilton's* work on kin selection* during the 1960s helped develop this new discipline.[10] Hamilton showed that closely related kin display more altruistic (selfless) behavior toward each other—a fact that is now known as Hamilton's rule.* Hamilton explained how altruistic behavior such as eusociality* in insects (that is, the existence of sterile worker classes) could have developed from kin selection.[11]

In Current Scholarship

The 1980s and 1990s saw a revival of structuralist* ideas in evolutionary biology; structuralism is an intellectual current, influential in fields such as anthropology and the study of language, in which (very roughly) different components of a language or a culture, say, are considered to be part of a system. This was partly due to the work of the biologists Brian Goodwin and Stuart Kauffman, who emphasized the contribution of self-organization to the course of evolution.[12] (Self-organization is when the components of a system interact to become ordered, in a previously disorganized system.)

From the 1980s onward, new data began to accumulate, leading scientists to understand that it is not different sets of proteins that control the way animals develop their phenotype* (visible characteristics). Instead, it is changes in the distribution of a small set of proteins that are common to all animals.[13] These proteins are called the *developmental-genetic toolkit.*[14] This knowledge had a great impact on the discipline of phylogenetics* (the study of the evolutionary relationships between a group of species), paleontology* (the study of fossilized remains) and comparative developmental biology,* and gave rise to a new discipline—evolutionary developmental biology, also known as evolutionary biology.[15]

Modern biologists are less concerned than Darwin had been with whether natural selection explains adaptation, because only 5 percent of genetic change is adaptive. As a consequence of the discovery of DNA*—something, of course, Darwin knew nothing about—they place much emphasis on random evolutionary change. (DNA is the material, carried in genes, that provides instructions for the growth and functioning of all living organisms.)

It is now accepted that two main processes cause evolutionary change: natural selection and random genetic drift (the process by which genetic information changes over time as animals reproduce). Evolution is not only driven by natural selection, as Darwin argued. It can also happen by chance if there are two equally good versions of a gene or allele* (gene variant) and one is luckier than the other, over generations, in spreading through a population. Genetic drift is random and therefore can be influenced by environmental occurrences such as natural catastrophes, etc.

To give a simple illustration, a child inherits the gene code for black hair from its mother and a second gene code for red hair from its father (the genotype). This individual then has a child. It can only pass on one copy of the gene and, by chance, passes on to its child the gene that codes for black hair. Over generations, if this scenario is replayed, the black hair color becomes dominant in the population and the code for red hair gets lost. Darwin, however, only knew about the observable form of organisms (the phenotype), and was more concerned with the evolution of these features.[16]

NOTES

1   Friedrich Leopold August Weismann, *Die Entstehung der Sexualzellen bei den Hydromedusen: Zugleich ein Beitrag zur Kenntniss des Baues und der Lebenserscheinungen dieser Gruppe* (Jena: Fischer, 1883).

2   Alfred Russel Wallace, *Darwinism: An Exposition of the Theory of Natural Selection, with Some of Its Applications* (London: Macmillan & Co, 1889).

3   Edward J. Larson, *Evolution: The Remarkable History of a Scientific Theory* (New York: Modern Library, 2004).

4   Michel Durinx and Tom J. M. Van Dooren, "Assortative Mate Choice and Dominance Modification: Alternative Ways of Removing Heterozygote Disadvantage," *Evolution* 63, no.2 (2009): 334–52.

5   J. Ollerton, "Speciation: Flowering time and the Wallace Effect," *Heredity* 95, no. 3 (2005): 181–2.

6   Larson, *Evolution*.

7   Larson, *Evolution*.

8   Thomas Hunt Morgan, *et al.*, *The Mechanism of Mendelian Heredity* (New York: Henry Holt and Company, 1915).

9   Larson, *Evolution*.

10 Joel L. Sachs, "Cooperation within and among species," *Journal of Evolutionary Biology* 19, no. 5 (2006): 1415–8.

11 Martin A. Nowak, "Five rules for the evolution of cooperation," *Science* 314 (2006): 1560–3.

12 Peter Corning, *Holistic Darwinism: Synergy, Cybernetics, and the Bioeconomics of Evolution* (Chicago: University of Chicago Press, 2010), 95–99.

13 John R. True and Sean B. Carroll, "Gene Co-option in Physiological and Morphological Evolution," *Annual Review of Cell and Developmental Biology* 18 (2002): 53–80.

14 Cristian Cañestro, Hayato Yokoi, and John H. Postlethwait, "Evolutionary Developmental Biology and Genomics," *Nature Reviews Genetics* 8, no. 12 (2007): 932–42.

15 Jaume Baguñà and Jordi Garcia-Fernàndez, "Evo-Devo: the Long and Winding Road," *International Journal of Developmental Biology* 47, nos. 7–8 (2003): 705–13.

16 Mark Ridley, *How to Read Darwin* (London: Granta Books, 2006).

MODULE 11
IMPACT AND INFLUENCE TODAY

KEY POINTS

- *On the Origin of Species* remains as relevant today as it was upon publication.

- Although most modern biologists operate in a paradigm of nature-and-nurture where both influence each other in a difficult-to-separate systems loop, the debates between hardwired behavior (what Darwin termed "instinct") versus social constructivism* (the idea that behavior is often constructed by surrounding culture) still flourish, as does the question of how unique humans are among animals.

- Fervent opposition to the theory of evolution still exists among some religious groups, who fear it leads to atheism.

Position

Inspired by Charles Darwin's theory of natural selection as set out in *On the Origin of Species by Means of Natural Selection*, a new discipline—sociobiology—arose that attempted to apply an evolutionary framework to social conditions.

It has been argued that some human behaviors are programmed in our genes from birth—a point of view termed "biological determinism."* Sociobiological theories have been spearheaded by biologists such as Richard Dawkins* and Edward Osborne Wilson,* among others. In *The Selfish Gene* (1976), Dawkins proposes that we cannot escape what is dictated by our genes, no matter how hard we try.[1] Expanding this idea and applying it more broadly to society, Wilson, writing in 1975, said that all societies "no matter how egalitarian, would always give a disproportionate share of power to men because of the fixed genetic differences between men and women."[2]

> ❝ You could give Aristotle a tutorial. And you could thrill him to the core of his being. Yet not only can you know more than him about the world. You also can have a deeper understanding of how everything works. Such is the privilege of living after Newton, Darwin, Einstein, Planck, and their colleagues. I'm not saying you're more intelligent than Aristotle, or wiser. For all I know, Aristotle's the cleverest person who ever lived. That's not the point. The point is only that science is cumulative, and we live later. ❞
>
> Richard Dawkins, "Science, Delusion and the Appetite for Wonder"

Evolutionary biologists* such as Stephen Jay Gould and Richard Lewontin* have criticized this approach as a reductionist* interpretation (an oversimplification) of human behavior. Recent sociological and scientific studies have shown that the determination of gender differences are trends, not defined behaviors.[3,4] Hard-line biological determinists,* on the other hand, such as the evolutionary psychologist* Steven Pinker,* define behavior as "male" or "female." Speaking at the Tanner Lectures in 1982, Lewontin offered an insight into why determinist ideas have nevertheless proliferated: "What makes these various inequalities between individuals, races, nations, and the sexes so problematic for us is the obvious contradiction between the fact of inequality and the ideology of equality on which our society is supposedly built."[5] In the past, an unequal society was justified as being the way God meant it to be. Some scientists – particularly evolutionary psychologists – now attempt to justify social inequalities using arguments from biology, claiming that inequality such as that between men and women is due to the inherent differences in the biology of the brain or via differently evolved behaviours. These contemporary scientists have been criticized for reinforcing dominant ideologies by using science to answer questions of a personal, social, or political nature by reasoning backwards with "Just So Stories"*.[6]

Interaction

Natural selection suggests that human beings are the product of a long evolutionary process, as they are specialized animals among many others. In the nineteenth century, this view was controversial because it challenged the biblical doctrine of human exceptionalism (also called human uniqueness),* which claims that humans are the pinnacle of God's creation. The idea created an identity crisis in Victorian society because many people refused to believe they could be so closely related to animals.[7]

Today, the issue of human uniqueness is debated just as widely as it was when the work was published in 1859.[8] The anthropologist Kim Hill* argues that what sets us apart from other animals is our dependence on culture and cooperation. The evolutionary psychologists Josep Call* and Michael Tomasello* have shown that great apes are able to determine the intentions of others and, in the unique case of humans, the ability to *share* intentions.[9] Tomasello concludes that this demonstrates the cognitive chasm between us and the other apes.[10]

However, other animals – such as the star-nosed mole, to take one example– [11] are equally specialized in very different ways. As Darwin himself put it, "the difference in mind between man and the higher animals, great as it is, certainly is one of degree and not of kind."[12] These, undoubtedly, are still the dominant views opposing claims of human exceptionalism. Theologians* such as Mark Harris* argue that scientific evidence used to back up claims of human uniqueness are only examples of the qualitative differences between ourselves and the other animals, because there is no one trait that sets humans apart from other primates.[13] Indeed, research into human origins has revealed more similarities with our biological cousins than it has differences, including development of the prefrontal cortex and the presence of mirror neurons.[14]

The Continuing Debate

Conversely, the idea of evolution has been uncontroversial in mainstream science for almost a hundred years. Since the 1930s and 1940s, when modern evolutionary theory incorporated genetic* science into Darwinian theory, most denials of the idea of evolution have originated from fundamentalist (literalist*) religious groups who maintain their belief in the creation myth. Arguments against the theory of evolution include objections to evidence, scientific methods, morality, and plausibility.[15]

A central text of the intelligent design movement is the book *Darwin on Trial* (1991) written by the lawyer Philip Johnson.* He uses a legal framework to structure his argument. To illustrate, the legal term "beyond a shadow of a doubt" (used in the rare cases when it is deemed there is sufficient evidence to say that something is true) is used to destroy a scientific theory.[16] However, the evolutionary biologist Stephen Jay Gould points out that a legal argument such as this is simply inappropriate when applied to science, because "science is not a discipline that claims to establish certainty."[17] Indeed, unlike religion, science was developed as a technique explicitly used to question itself, a tool known as the scientific method*, where the tester observes phenomena, questions, hypothesises causality, experiments and analyses.

Certainly, the belief that evolution promotes atheism has produced vehement opposition to it.[18] Creationists claim that proponents of evolutionary theory are atheists who cannot see beyond material causes and facts, to the detriment of a complete understanding of the nature of existence.[19] However, such claims are tenuous; a poll conducted in 2014 found that 40 percent of scientists in the United States believe in a god—similar to results conducted among the general American public.[20,21]

Religious literalists support the creation myth unswervingly, but there is evidence to suggest that the biblical book of Genesis is a retelling of earlier narratives—encompassing themes from the ancient Middle Eastern myth of Gilgamesh* and ancient Mesopotamian* understandings of the world. One Mesopotamian myth describes how the goddess Ninti was sent to heal another god called Enki, whose body had been ravaged by disease. The Sumerian word *Ninti* has a double meaning—both "rib" and "life"—a theme repeated in Genesis, in which Eve is created from Adam's rib.[22]

Indeed, the Christian religious scholars Bruce Waltke* and Conrad Hyers* caution against a literal interpretation of the creation myth on the grounds that it is aligned with these tenets of Mesopotamian science and religion, arguing against such a reading precisely because it leads to the denial of evolution. For them, current scientific knowledge should be incorporated into the creation narrative instead.[23] The Roman Catholic Church has done precisely that, reconciling its belief in a deity with evolution by advocating theistic evolution* (the idea that evolution is a process put into action and guided by the hand of God). Prominent Hindu, Sikh, Buddhist and Jewish scholars have done similarly. [24] Mainstream Islam, on the other hand, has not, although branches of Islam, such as the reformist Ahmadiyya movement, accept Darwinian principles with the proviso of God directing the process,[25] more like a "game of chess" than a "game of chance."[26]

NOTES

1   Richard Dawkins, *The Selfish Gene* (Oxford: Oxford University Press, 1990).

2   E.O. Wilson, "Human Decency Is Animal," *New York Times Magazine*, October 12, 1975.

3   B. J. Carothers and H. T. Reis, "Men and Women Are From Earth: Examining the Latent Structure of Gender," *Journal of Personality and Social*

Psychology 104, no. 2 (2013): 385–407.

4   Daphna Joel *et al.*, "Sex Beyond the Genitalia: The Human Brain Mosaic,"
    PNAS 112, no. 50 (2015): 15468–73.

5   R. C. Lewontin, "Biological Determinism," The Tanner Lectures on Human
    Values, University of Utah, March 31, April 1 1982. Accessed February 5,
    2016, http://tannerlectures.utah.edu/_documents/a-to-z/l/lewontin83.pdf.

6   Lewontin, "Biological Determinism," The Tanner Lectures On Human Values.

7   Edward J. Larson, *Evolution: The Remarkable History of a Scientific Theory*
    (New York: Modern Library, 2004).

8   H. Guldberg, "Restating the Case for Human Uniqueness," *Psychology
    Today*, November 8, 2010.

9   J. Bräuer, J. Call, and M. Tomasello, "Chimpanzees really know what others
    can see in a competitive situation," *Animal Cognition* 10 (2007): 439–48.

10 M. Tomasello *et al.*, "Understanding and sharing intentions: the origins of
    cultural cognition," *Behavioral and Brain Sciences* 28, no. 5 (2005): 675–
    735.

11 Robert Barton, speaking at a 2011 London Evolutionary Research Network
    debate on the subject of whether human uniqueness exists at University
    College London.

12 Charles Darwin, *The Descent of Man, and Selection in Relation to Sex*
    (London: John Murray, 1872), 82.

13 Mark Harris, "Human uniqueness, and are humans the pinnacle of
    evolution?" *Science and Religion @ Edinburgh*, September 7, 2014,
    Available at: http://www.blogs.hss.ed.ac.uk/science-and-religion/2014/09/07/
    human-uniqueness-and-are-humans-the-pinnacle-of-evolution/. [Accessed
    February 16, 2016,]

14 J.B. Smaers *et al.* "Primate Prefrontal Cortex Evolution: Human Brains Are
    the Extreme of a Lateralized Ape Trend." *Brain, Behavior and Evolution* 77
    (2011): 67–78; P. F. Ferrari, L. Bonini, and L. Fogassi, "From monkey mirror
    neurons to primate behaviours: possible 'direct' and 'indirect' pathways,"
    *Philosophical Transactions of the Royal Society of London, Biological
    Sciences* 364, no.1528, (2009): 2311–2323.

15 Peter Cook, *Evolution Versus Intelligent Design: Why All the Fuss? The
    Arguments for Both Sides* (Australia: New Holland Publishing, 2007).

16 Phillip E. Johnson, *Darwin on Trial* (Downers Grove, IL: InterVarsity Press,
    1991).

17 S. J. Gould, "Impeaching a Self-Appointed Judge," *Scientific American* 267,

no. 1 (1992): 118–21.

18 Lee Strobel, *The Case for a Creator: A Journalist Investigates Scientific Evidence That Points Toward God* (Grand Rapids, MI: Zondervan, 2004).

19 Phillip E. Johnson, "The Church of Darwin," *Wall Street Journal,* August 16, 1999.

20 Larry Witham, "Many Scientists See God's Hand in Evolution," *Reports of the National Center for Science Education* 17, no. 6 (November–December 1997): 33.

21 Bruce A. Robinson, "Beliefs of the U.S. Public about Evolution and Creation," accessed 14 April 2015, www.*ReligiousTolerance.org*.

22 Samuel Henry Hooke. *Middle Eastern Mythology* (Dover Publications, 2013), 115.

23 Conrad Hyers, *The Meaning of Creation: Genesis and Modern Science* (Louisville: Westminster John Knox, 1984).

24 "Religious Groups: Opinions of Evolution". The Pew Forum on Religion and Public Life (2007); Hardev Singh Virk "The Origin of Life and Evolution according to Science and Gurbani," *The Sikh Review* 360, Sector 71, (January, 2008); Religious Differences on the Question of Evolution, Analysis," The Pew Forum on Religion and Public Life (4 February, 2009); Nathan Slifkin "The Challenge of Creation: Judaism's Encounter with Science, Cosmology, and Evolution," Brooklyn: ZooTorah/Lambda Press; section two, *Cosmology* (2010): 157–190; David J. Fine *Intelligent Design YK Day*, 1 Temple Israel 2009, accessed July 10, 2013. Temple Israel: synagogue.org/portals/0/rabbi sermons/yom kippur-intelligent design.pdf.

25 S. Hameed, "Bracing for Islamic creationism," *Science* 322/5908 (2008): 1637–8.

26 Mirza Tahir Ahmad *Revelation, Rationality, Knowledge and Truth* (Islam International Publications Ltd, 1998).

MODULE 12
WHERE NEXT?

KEY POINTS

* The theories found in *On the Origin of Species* have never been disproven and will inform scientific investigations far into the future.

* Today, emerging research into DNA and genetics is dependent on the theory of natural selection, and will probably continue to be so in the future.

* The theories put forward in *On the Origin of Species* form the basis for all contemporary biological sciences.

Potential

Charles Darwin's *On the Origin of Species by Means of Natural Selection* discusses the evolutionary* past of life on earth—but many people wonder what the evolutionary future may be for humans.

The Welsh geneticist Steve Jones argues that human evolution is slowing down, because we are no longer subject to natural selection—the *fittest* individuals, that is, no longer drive evolutionary change. In the nineteenth century, when Darwin published *On the Origin of Species*, fewer than half of British children survived to 35 years of age. Today that number is around 95 percent,[1] largely due to the advances made in medicine in the treatment of various illnesses and diseases. Now those individuals deemed weakest continue to survive and have children. In Jones's words, "Darwin's machine has lost its power."[2] This is a problematic argument because natural selection is neutral; here the fittest would mean those who have access to modern medicine, an environment that permits survival.

> **❝** It will be possible, through the detailed determination of amino-acid sequences of hemoglobin molecules and of other molecules too, to obtain much information about the course of the evolutionary process, and to illuminate the question of the origin of species. **❞**
>
> Linus Pauling, *Molecular Disease and Evolution*

The evolutionary psychologist Geoffrey Miller* also disagrees. Viral and bacterial pathogens can now, due to modern technology such as airplanes, spread more easily to various parts of the globe. Miller predicts that epidemics will become important in shaping the human immune system, and will result in future generations of humans possessing stronger immune systems.

Future Directions

The ancient Greek philosopher Aristotle envisaged life as a Great Chain of Being (*scala naturae*), with every organism occupying a place in a hierarchy, and there is still a "tendency in evolutionary discourse to describe life's history as a progression towards increasing complexity."[3] There are however, cases where simple organisms—such as gutless tapeworms or blind cave fish—have arisen from more complex ones. This is termed "reductive evolution", and is merely a case of simplicity here being more adaptive.

It has also been found that organisms can occasionally lose their ability to perform a function previously thought necessary to their survival, without it affecting their ability to survive or multiply. This was first discovered in the ocean-dwelling plankton *Prochlorococcus*—a common photosynthetic microorganism (an organism that can convert sunlight into energy). Researchers found that it had lost the gene that helps to neutralize hydrogen peroxide, a compound that can destroy cells.[4] Instead, it relies on other nearby microorganisms to eliminate hydrogen peroxide from its environment.

For microorganisms, carrying genes and manufacturing proteins requires a great deal of energy, so discarding certain genes allows them to live more efficiently. The researchers who first made this observation have termed it the Black Queen hypothesis*—the principle that natural selection drives microorganisms to lose essential functions when there is another species nearby to perform them.[5] Evolution can favor coexistence between helpers that perform the function and beneficiaries that require it. The fitness of helpers is not reduced, resulting in a stable coexistence that can lead to the evolution of mutualism,* where two organisms of different species cooperate and benefit from the arrangement.

While Darwin stressed competition and conflict, the Canadian biologist Brian Goodwin* argued that survival is a matter of finding a niche.* For Goodwin, organisms that survive are not better than those that have become extinct. Instead evolution is "like a dance" with organisms "simply exploring a space of possibilities."[6] This neutral aspect is likely accurate, as it avoids ascribing agency to natural selection, as those who point to the example of the blind cave fish as "going backward" would be in danger of doing.

Summary
While the idea of evolution preceded Darwin, it was his idea of natural selection that made evolution plausible: individuals vary, random selection makes some individuals better suited to their environment than others, individuals better adapted will have disproportionately more offspring. By making evolution the source of different species of animal life rather than God and eliminating a Divine Creator, scientific explanations for natural phenomena became paramount.[7] For this innovation, he was, and continues to be, celebrated.

Darwin founded the science of evolutionary biology, with natural selection as its underlying principle. Another of his contributions to this field was to suggest that species change over time, a process called gradualism. He also pictured evolution developing in branches, rather than progressing in a linear fashion, implying that all species descend from a single unique origin. While Darwin thought that selection takes place at the level of the individual, we now know it is at the level of the gene. This has recently given rise to a new field of medicine called gene therapy,* which makes it possible to treat genetic disorders, such as cystic fibrosis,* cancer, and certain infectious diseases (including HIV).* Doctors can perform such therapy in utero (in the womb), potentially treating a life-threatening disorder before a child is even born. While gene therapy could spare future generations from particular genetic disorders, it might also affect the development of a fetus in ways we cannot predict, or have long-term side effects that are as yet unknown.[8] Consequently it remains controversial.

Darwin's great idea of natural selection continues to evolve and adapt via the modern evolutionary synthesis of the 1930s and 1940s and into our present century. Darwin's own words regarding the envisioned future acceptance of his theory in *On the Origin of Species* — "In the distant future I see open fields for far more important researches,"[9] — may well have pleased him, for in its survival, and via its adaptations, natural selection has been proven astonishingly fit.

NOTES

1   Office of National Statistics, Mortality Rate in the UK 2010, available at http://www.ons.gov.uk/ons/rel/mortality-ageing/mortality-in-the-united-kingdom/mortality-in-the-united-kingdom–2010/mortality-in-the-uk-2010.html. [Accessed February 5, 2016,]

2   Steve Jones, speaking at a lecture marking the bicentenary of Darwin's birth and the 150th anniversary of *On the Origin of Species*, at the University of Cambridge.

3   Jeffrey J. Morris, Richard E. Lenski, and Erik R. Zinser, "The Black Queen Hypothesis: Evolution of Dependencies through Adaptive Gene Loss," *mBio* 3, no. 2 (2012): 1–7.

4   Morris, "The Black Queen Hypothesis," 1–7.

5   Morris, "The Black Queen Hypothesis," 1–7.

6   Brian Goodwin, *How the Leopard Changed Its Spots: The Evolution of Complexity* (Princeton New Jersey: Princeton University Press, 2001), 98.

7   Ernst Mayr, "Darwin's Influence on Modern Thought," *Proceedings of the American Philosophical Society* 139, no. 4 (Dec 1995): 317–25.

8   For a review on the issues surrounding the application of gene therapy, see Sonia Y. Hunt, "Controversies in Treatment Approaches: Gene Therapy, IVF, Stem Cells, and Pharmacogenomics," *Nature Education* 1, no. 1 (2008): 222.

9   Charles Darwin, *On the Origin of Species by Means of Natural Selection, or the Preservation of Favoured Races in the Struggle for Life*. Introduction and Notes by Gillian Beer (Oxford: Oxford University Press (1996, 2008): 369.

GLOSSARY

GLOSSARY OF TERMS

Abolitionism: a movement to end slavery in western Europe and the Americas.

Allele: a variant form of a gene. Some genes come in a variety of different forms.

Altruism: concern for or devotion to the welfare of others. Behavior that is disadvantageous to the person acting, but beneficial to the recipient, is termed altruistic.

Anatomist: a scientist who specializes in comparing and contrasting the anatomies of different species.

Anthropology: the study of human beings, including their cultures, social lives and evolutionary history. Cultural (or sociocultural) anthropology traditionally is more associated with the two former arenas, and physical (or biological) anthropology with the latter.

Arthropod: the name given to a group of animals without a backbone, but with a segmented body and jointed limbs. Insects, millipedes, crustaceans, and spiders are all arthropods.

Atheism: the belief that no god or divine being exists.

***Beagle*:** the HMS ("Her/His Majesty's Ship") *Beagle* was a Royal Navy vessel on which Charles Darwin spent five years sailing around the world on a scientific voyage, collecting specimens and developing his theory of natural selection.

Behavioral ecology: a field that uses evolutionary theory and the environment in which an animal lives (including factors such as

81

predators and the weather) to understand behavior.

Biological anthropology: also referred to as "physical anthropology," this is the biological study of human beings as animals.

Biological determinism: the suggestion that all (or most) human behavior is dictated by genes, rather than by upbringing or personal choice.

Biological inheritance: the various characteristics that can be passed from parent to child.

Black Queen hypothesis: the idea that microorganisms sometimes lose the ability to perform a function that is essential for their survival if other microbes in their immediate environment can perform the function for them. This adaptation encourages microorganisms to live in cooperative communities.

Botany: the study of plants.

Colonialism: acquiring full or partial political control over another territory or country, resulting in the exploitation of its peoples and resources.

Comparative anatomy: the study of the comparison and contrast of the anatomies of different species.

Comparative developmental biology: a field of inquiry using natural variation and disparity to understand the patterns of growth of life forms at all levels.

Creationism: the belief that life and the natural world were created by a divinity or divinities.

Cystic fibrosis: a genetic disorder that compromises the function of organs that are fundamental to life.

Descent with modification: a term describing how over time and generations, the traits conferring reproductive advantage become more common within a population.

Divergence: When a species splits to become two or more separate species.

DNA: an abbreviation for *Deoxyribonucleic acid*, a molecule that carries most of the genetic instructions used in the development, functioning, and reproduction of all living organisms and many viruses.

Ecological niche: the ecological role of an organism in a community. It applies particularly to food consumption.

Enlightenment: a period and intellectual movement in Western Europe during the seventeenth and eighteenth centuries. Reason was emphasized over tradition and ideas of divine interference in the affairs of humans. Prominent scholars of the period included the French authors Voltaire and Rousseau, and the German philosopher Immanuel Kant.

Embryology: the study of the development of the embryonic stages of animals.

Erosion: the geological process—driven by wind, sun, or water—that breaks down soil or rock structures.

Essentialism: the view that for any specific entity (such as an animal or a physical object) there is a set of attributes (usually hidden or unseen) that are necessary to its identity and function. The concept arose from the work of the Greek philosophers Plato and Aristotle.

Eugenics: the concept that one can improve a population of organisms by controlled breeding.

Eusociality: a term describing organisms that live in a cooperative group in which usually only one female and several males are reproductively active, the rest being (usually closely related) nonbreeding individuals who care for the young, or protect and provide for the group. Examples include termites, ants, and naked mole rats.

Evolution: the generational changes in heritable traits in organisms.

Evolutionary biology: the study of species change over time. Subfields in this area include taxonomy, ecology, population genetics, and paleontology.

Evolutionary psychology: the study of psychology in an evolutionary context.

Evolutionism: the idea that societies start out as "primitive" and advance toward "civilization."

Extinction: a biological term for the end of an organism or species.

Gene: a part of DNA that encodes a functional protein product. It is the molecular unit of heredity.

Gene therapy: the treatment of certain medical disorders caused by genetic anomalies or deficiencies, by altering or replacing the genes in a patient's cells.

Genesis: the first book of the Pentateuch (Genesis, Exodus, Leviticus, Numbers, Deuteronomy), part of the Jewish and Christian scriptures. The book describes the creation of the earth and humans, and the expansion of the human race, along with the story of Abraham and his descendants.

Genetic drift: nonadaptive change that is caused by gene variation due to purely random sampling. For example, the frequency of green-eyed alleles in a particular population could be reduced due to a chance environmental disaster.

Genetics: the study of genes and genetic variation in life forms.

Genotype: the genetic blueprint for an individual organism. Genotypic variation is variation due to genetic differences.

Geography: the study of the planet's land and environments and the ways in which humans interact with their ecological niches and environments.

Geology: the study of the physical earth, or any celestial body, including fields as diverse as plate tectonics or evolution.

Gilgamesh: *The Epic of Gilgamesh* is a Mesopotamian poem widely considered to be one of the earliest surviving works of literature. The epic concerns the story of the demigod king Gilgamesh, not least his quest for immortality.

Gradualism: in evolutionary studies, the concept that species change in intermediate stages over time periods, and generationally.

Great apes: Orangutans, gorillas, chimpanzees, bonobos and humans.

Great Chain of Being (*scala naturae*): a concept first put forward by the ancient Greek philosopher Plato. The main idea is that organisms all belong somewhere along a hierarchy. So everything has its "place," because that is the way God intended it to be.

Great Flood: Noah is the principal character mentioned in the biblical book of Genesis in the story of Noah's ark, which tells how God destroyed all living things on earth by way of a flood; two of each animal were saved on Noah's ship, or ark.

Great Reform Act: legislation enacted in England and Wales in 1832, in the face of fierce opposition, that gave groups of previously disenfranchised men the right to vote. Similar legislation enacted the same year did the same for Scotland and Ireland.

Group selection: the process by which an individual acts for the good of the group or species, a theory elaborated upon in 1986 by the English zoologist V. C. Wynne-Edwards that has since been discredited. The concept was first championed by scientists such as the Nobel prize-winning zoologist Konrad Lorenz.

Hamilton's Rule: explains the conditions required for altruism to evolve, for example when a creature helps its own close relatives to survive at the cost of its own wellbeing. The formula is $r \times B > C$, where r is the degree of relatedness between individuals, B is the benefit to the recipient, and C is the cost for performing that particular act.

HIV: stands for human immune deficiency virus. This is a retrovirus. The immune system of individuals who are infected has a reduced function. HIV is a cause of AIDS.

Human uniqueness/human exceptionalism: the belief that humans are an exceptional and unique species compared to all other animals. Exceptionalists usually reference Abrahamic religious beliefs, such as faith in the existence of the human soul, as a justification for creationism.

Hybridism: the offspring of two different species, subspecies, or occasionally even genera.

Intellectual priority: the acknowledgment that a seminal work is the first of its kind and represents a major step forward.

Intelligent design: a term used to describe the idea that the earth was designed and created with purpose by an individual god or gods.

Intersexual selection: the process that occurs when different sexes are attracted to each other, often due to particular characteristics. The sex with the more expensive gametes (usually female) tends to be choosier, as they have more to lose. The process is therefore often called "female choice".

Intrasexual selection: the process that occurs when members of the same sex (most often males) compete with each other for opportunities to mate.

Just So Stories: A popular set of fantastical "origin" stories for animals, written by Rudyard Kipling and first published in 1902. The Just So Stories purport to explain how animals – among them the kangaroo, the armadillo and the rhinoceros – acquired their main characteristics, but they are humorous and moralistic rather than biologically accurate.

Kin selection: the way certain behaviors are favored over evolutionary time between individuals that are closely related.

Lamarckism: term in Darwin's time for the transmutation (changing) of species, which went against the official scientific understanding that only fixed states of organisms exist. These days it generally means the incorrect idea that parents develop traits in their lifetime that their offspring will inherit.

Liberalism: the idea that the government should be responsible for protecting the freedom and equality of individuals in a society.

Life sciences: the fields of science that deal with living organisms— neuroscience, botany, and virology, for example.

Linguistics: the study of the structures and nature of language.

Linnean Society: a society founded in London to promote the study of natural history and taxonomy (the categorization of organisms).

Literalism: belief in the literal truth of the Bible and strict adherence to its teachings.

Malthusian catastrophe: a hypothetical scenario whereby all the members of a society are forced to return to self-sufficient farming in order to provide for their families. This is predicted to happen when population growth has overtaken agricultural production.

Mammal: the name for a group of animals that have a backbone and where the young are fed by milk produced in the mammary glands. Mammals include dolphins, humans, dogs, and cows, to name a few.

Materialists: in this sense, those who look for material causes alone for all material phenomena.

Mesopotamian science: all forms of scholarly inquiry into natural and cultural phenomena, both real and imagined, in the region of ancient Mesopotamia. This includes the invention of writing by the Sumerians, the division of time into the 60-second minute and the 60-minute hour, and perhaps invention of the wheel.

Modern evolutionary biology/modern evolutionary theory/ new synthesis: a reworking of Darwinian evolutionary theory in the 1930s and 1940s to incorporate genetics and other newer theories, such as kin selection and population demographics.

Monsters: Darwin called mutations "monsters." Before the work of the pioneering geneticist Gregor Mendel became widely known, there was no clear understanding of genetics and inheritance.

Mutation: a change in the sequence of an organism's genome—that is, in its heritable genetic material.

Mutation theory: the idea that new species are formed not by continuous variations, as Darwin suggested, but by sudden variations, called mutations. First proposed by the biologist Hugo de Vries in 1901, it stated that mutations are inherited through successive generations.

Mutualism: where two organisms of different species cooperate, with each benefitting from the arrangement.

Natural selection: a mechanism to explain species change. Charles Darwin and the naturalist Alfred Russel Wallace proposed that organisms over-reproduce; they strive for survival in varying environments; those with traits most favorable to survival live to reproduce and pass on those traits to their offspring, while those species that do not survive long enough to reproduce become extinct.

Naturalist/natural historian: a scholar of the natural world.

Naturalistic fallacy: that belief that because something is "natural" it is also "good."

Nazis: a contraction referring to members and supporters of the NSDAP, the far right-wing German political party led by Adolf Hitler. Nazi policies included eugenics and belief in the existence of a "master race" possessed of superior genetic characteristics.

Neo-Darwinism: a period in time when scientists, namely Wallace and Weismann, came to reject Lamarckian inheritance and promoted natural selection instead as an explanation for how evolution works. The term was coined by George Romanes in 1895.

Nonconformists: also known as Unitarians, this is a seventeenth-century term for people belonging to English Protestant religions that were not the official Church of England (notably Puritans and Methodists).

Oceanography: the scientific study of oceans.

Orthogenesis: the biologist Theodor Eimer's theory that species change as a direct result of some internal force within an individual, which seeks to modify the current type.

Paleontology: the scientific study of animals and plants that existed in the past and are now preserved in rocks. Scientists working in this field collaborate closely with geologists to identify the period of time when both rock and the fossilized remains of the organism existed.

Pangenesis: Darwin's theory on inheritance, that has since been proven to be false, whereby information from each cell in the body of an organism travels (in the form of particles) to the reproductive organs; in the reproductive organs all this information is merged and goes to make the sperm and egg.

Paradigm: an accepted mode of thinking or doing something in a particular way.

Parent species: the species that gives rise to other species, usually over long time periods. It usually means the most recent common ancestor of two different species.

Phenotype: the visible characteristics of an individual. The individuals' phenotype is shaped by not only its genes (or its genetic makeup) but also by the environment in which it grows up and lives.

Phylogenetics: study of the evolutionary relationships between species.

Plinian Society: A private members' club formed by students at the University of Edinburgh, for the purpose of reading and discussing articles published by scientists on issues to do with natural history.

Positivism: the principle that knowledge can only advance through the analysis of scientifically verifiable facts.

Primatology: the study of the order of primates, to which human beings and the other great apes belong.

Punctuated equilibrium: a theory of evolutionary biology that suggests evolution is not an incremental process, as argued by Darwin, but rather comprises short periods of rapid change, followed by longer ones of stasis.

Reductionism: different philosophical positions or theories connected to each other reduced to a "simpler" or more "basic" form.

Relativism: the idea that no absolute truth exists, and that truth is relative. This is often applied in sociocultural anthropology, which suggests that human behavior and beliefs should be understood in the context of their particular culture.

Religious fundamentalism: see literalism.

Scientific method: the systematic, logical, and reproducible investigation of the universe, with the aim of producing and testing hypotheses, and ultimately of constructing an accurate and detailed representation of how it works.

Scientific priority: a general term referring to work seminal in a particular area and regarded as a major advancement, such as the discovery of the double helix as the structure of DNA.

Social constructivism: the idea that behavior is often constructed by the surrounding culture.

Sociobiology: the field of science investigating the evolutionary origins and function of human social behavior.

Sociocultural anthropology: the study of human cultural variability, in the present or over particular time periods.

Sexual selection: the theory that males compete against each other for access to females, and that females (and to a lesser extent males) choose those with whom they wish to mate. As a result, some traits deemed "attractive," such as the peacock's tail, may be passed on to future generations in greater numbers than other traits.

Social Darwinism: the attempted application of Darwin's mechanism of natural selection to social structures. The implication is that certain races are more "evolved" than others, and that "successful" societies are the fittest, with greater rewards for the strong or most competitive.

Speciation: a process whereby a new species evolves. This occurs when a single species splits into two or more independent lineages.

Species: the largest group of organisms in which two individuals are capable of producing fertile offspring, typically using sexual reproduction.

Structuralism: a theoretical framework used to understand human culture: the patterns and interconnectedness of human interactions are sketched, and the resulting pattern or structure is used to better understand human culture.

Subspecies: a group of organisms that can interbreed, and sometimes do so in the wild. In taxonomy, species come after subspecies. This is the first stage of speciation—when new species are formed.

Survival of the fittest: this term is used to refer to individuals that are best equipped to survive and produce offspring. Thus, in biology, the word "fittest" refers to those individuals that are able to produce the greatest number of surviving fertile offspring during their lifetime.

Taxonomy: in biology, the systematic classification of living things. Taxonomy and nomenclature (the scientific naming of organisms) was revolutionized by the Swedish naturalist Carl Linnaeus in 1735.

Temperance movement: a series of mass social and political movements to outlaw or reduce the legality and drinking of alcohol, which reached its peak in terms of political clout in Europe and North America from the mid-eighteenth century to the early twentieth century.

Theistic evolution: the idea that evolution is a process put into action and guided by the hand of God.

Theology: the systematic study of religious ideas, commonly conducted through religious scripture.

Transmutation (of species): a phrase coined by the French biologist Jean-Baptiste Lamarck in 1809 to describe the possibility of species change from uniform prototypes.

Uniformitarianism: the idea that the same forces that change the physical earth today also operated in the past; uniformitarianism was first proposed by the Scottish geologist James Hutton in 1788.

PEOPLE MENTIONED IN THE TEXT

Al-Jahiz (776-868/9 CE) was an Arab writer, religious thinker and polymath, born in what is now Iraq. He was author of the *Book of Animals*, a multi-volume encylopaedia in which he observed: "Animals engage in a struggle for existence, and for resources, to avoid being eaten, and to breed."

Al-Tusi, Nasir al-Din (1201-1274 CE) was a Persian scientist, born in what is now Iran shortly before the Mongol conquests. Appointed as an advisor to the Mongol khan Hulegu, al-Tusi was responsible for the construction of a new observatory at Baghdad and was the author of numerous works on astronomy as well as commentaries on older Greek mathematical and scientific texts.

Anaximander (c.610-546 BCE was the earliest Greek philosopher of whose writings some trace survives. He speculated about the nature of the universe, considering what he termed "the boundless" to be the origin of all things.

Sir Charles Bell (1774–1842) was a Scottish anatomist noted for identifying the function of sensory and motor nerves. He also wrote on religion, philosophy, and theology.

Janet Browne (b. 1950) is a British historian of science and one of the most acclaimed biographers of Charles Darwin, known for her works *Charles Darwin: Vol. 1, Voyaging* and *Charles Darwin: Vol. 2, The Power of Place*.

Josep Call (b. 1967) is a Spanish comparative psychologist. His research focuses on investigating nonhuman primate cognition and comparing it to human intelligence. He is currently the director of the Wolfgang Köhler Primate Research Center at the Max Planck Institute, Leipzig, Germany.

Robert Chambers (1802–71) was a Scottish scientist and publisher. His pre-Darwinian evolutionary ideas were published anonymously in 1844 in a work called *Vestiges of the Natural History of Creation*.

Carl Erich Correns (1864–1933) was a German geneticist and botanist. He is best known for discovering the principle of biological heredity, and for rediscovering the work of Gregor Mendel on pea plants and inheritance. He achieved this at the same time as Hugo de Vries, but independently.

Georges Cuvier (1769–1832) was a French naturalist and zoologist who championed the fixity of species on both religious and scientific grounds. He was best known for *Tableau elementaire de l'histoire naturelle des animaux* (1798).

Emma Darwin (1808–96), born Emma Wedgwood, became Charles Darwin's wife and helped him raise nine children. She was Darwin's first cousin and they both belonged to the notable Wedgwood family, makers of Wedgwood pottery.

Erasmus Darwin (1731–1802) was grandfather to Charles Darwin. A prominent scientist and physician, he had himself promoted ideas about the transmutation of species in the form of poetry.

Robert Waring Darwin (1766–1848) was an English medical doctor, and father of the naturalist Charles Darwin.

Richard Dawkins (b. 1941) is a science writer, evolutionary biologist, and prominent atheist. He is best known for his 1976 book *The Selfish Gene*, which argues that natural selection takes place at gene level.

Democritus (460 BCE–70 BCE) was an ancient Greek pre-Socratic philosopher. He is best known for suggesting the atomic theory of the universe. He is considered a materialist, believing everything to be the result of natural laws.

Hugo de Vries (1848–1935) was a Dutch botanist and geneticist. In the 1890s he came up with the concept of genes, after rediscovering the work of Gregor Mendel on heredity in pea plants. He is also notable for introducing the term mutation and developing a theory of evolution based on mutation occurring in genes.

Theodosius Grygorovych Dobzhansky (1900–75) was a prominent Ukrainian-born geneticist and evolutionary biologist. He later moved to work in the United States. He was a central figure in the field of evolutionary biology and unifying modern evolutionary synthesis.

Niles Eldredge (b.1943) is an American paleontologist and curator at the American Museum of Natural History. He is best known for proposing, with Stephen Jay Gould, the theory of punctuated evolution, which challenges Darwin's view that evolution occurs gradually.

Empedocles (c.492–432 BCE) was a Greek pre-socratic philosopher, poet and follower of Pythagoras whose work now exists only in fragments. He is often credited as the author of the four-element theory of matter, which suggests that everything in the world is composed some combination of air, earth, fire and water.

Gustav Heinrich Theodor Eimer (1843–98) was a German zoologist. He is credited with popularizing the term orthogenesis to describe evolution guided in a specific direction.

Charles Sutherland Elton (1900–91) was an English zoologist and ecologist best known for his studies on modern population ecology.

Ronald Fisher (1890–1962) was a British statistician who incorporated Mendelian genetics into Darwinian theory, resulting in what is now known as the modern evolutionary synthesis. He is one of the key founders of what was later known as population genetics. Less favorably, he was also a prominent eugenicist.

Robert FitzRoy (1805-1865) was a British Royal Navy officer, scientist, and later Governor of New Zealand. He is best known for his captaincy of HMS *Beagle,* the ship which carried Charles Darwin around the world.

Francis Galton (1822–1911) was an English anthropologist and statistician, whose achievements included the invention of fingerprint identification. He was also a prominent eugenicist, and the half-cousin of Charles Darwin.

Brian Carey Goodwin (1931–2009) was a Canadian mathematician and biologist. He was one of the founders of theoretical biology—a branch of mathematical biology that uses methods from mathematics and physics to understand processes in biology.

John Gould (1804-1881) was a British ornithologist and taxidermist. Born in Dorset, the son of a gardener, he is noted for his part-work publication of illustrated guides to the birds of the Himalayas, Europe & Australia.

Stephen Jay Gould (1941–2002) was an American paleontologist, evolutionary biologist, and popular science writer who opposed sociobiological determinism.

Robert Edmond Grant (1793–1874) was a Scottish doctor and anatomist who taught Darwin at Edinburgh University. Grant later moved to London where he set up the now famous Grant museum of zoology and became the first professor of zoology in Britain. His research on marine invertebrates established that sponges are in fact animals.

William (Bill) D. Hamilton (1936–2000) was an English evolutionary biologist who theorized that social favoritism shown toward indirect kin (nieces, for example) often appears as altruism when in fact it serves the individual's fitness to favor kin. He is best known for Hamilton's Rule. Hamilton was a forerunner of gene-based evolutionary theorizing from scientists such as Richard Dawkins.

Mark Harris is a Senior Lecturer in Science and Religion at the University of Edinburgh. He is best known for his 2013 book *The Nature of Creation*, which explores the links between science and religion.

John Stevens Henslow (1796–1861) was a British botanist, clergyman, and geologist. At the University of Cambridge he introduced a teaching technique that fostered independent discovery, and became a source of inspiration to the young Charles Darwin.

Kim Hill is an American anthropologist. His work focuses on the evolutionary ecology of human behavior. He is known for his views on human uniqueness, linking humans' success to their sociality.

Joseph Dalton Hooker (1817–1911) was a well-regarded botanist and explorer, director of the Royal Botanic Gardens, Kew, and close personal friend of Darwin.

James Hutton (1726–97) was a Scottish geologist best known for his theory of uniformitarianism.

Thomas Henry Huxley (1825–95) was an anatomist, widely known as "Darwin's bulldog" for his well-publicized defense of Darwin's ideas in *On the Origin of Species by Means of Natural Selection*. He wrote *Evidence as to Man's Place in Nature* (1863), where he explicitly stated that humans and apes descended from a common ancestor and that humans, like other animals, evolved and thus were and are subject to natural selection.

Conrad Hyers (1933-2013) was an American Presbyterian minister and Professor of Religion at Gustavus Adolphus College in Minnesota. His work cautions against literal acceptance of the creation myth.

Fleeming Jenkin (1833–1885) was a British engineer noted for inventing the cable car.

Phillip Johnson (b. 1940) is a former University of California, Berkeley, law professor. He is known as the father of the intelligent design movement, the theory that argues that life could not have developed by chance, and must therefore be the design of an intelligent being.

Stephen Jones (b. 1944) is a Welsh geneticist and evolutionary biologist at University College London.

Stuart Alan Kauffman (b.1939) is an American theoretical scientist and trained medical doctor. He is best known for arguing that the complexity of biological systems and organisms might result as much from self-organization as from Darwinian natural selection.

Motoo Kimura (1924–94) was a Japanese biologist who developed the neutral theory of molecular evolution, which suggests that it is drift and not natural selection that is the cause of most species change.

Reverend Charles Kingsley (1819–75) was an English cleric who vocally spoke and wrote in favor of Darwinian ideas and evolution.

Astrid Kodrick-Brown is an American ecologist and evolutionary biologist who focuses on the behavior of freshwater fish, especially the evolution of mate recognition systems and their role in speciation.

Jean-Baptiste Lamarck (1744–1829) was a French biologist who suggested the concept of changeable species in his 1809 work *Philosophie Zoologique,* a work that also stated that changes can be inherited generationally.

Georges-Louis Leclerc, Comte de Buffon (1707–88) was a French naturalist, mathematician, and director of the Jardin du Roi (later called Jardin des Plantes), a notable botanical garden in Paris. He is regarded as the father of natural history during the Enlightenment period. His best-known work, *Histoire Naturelle* (1749–1788) is a 36-volume encyclopedia of the animal and mineral kingdoms.

Richard Lewontin (b. 1929) is an evolutionary biologist and geneticist best known for his book *Not in Our Genes: Biology, Ideology and Human Nature,* written with Steven Rose and Leon J. Kamin.

Carl Linnaeus (1707–1778) was the Swedish botanist who devised a system for sorting all known organisms, including human beings, into subsets. In his 1735 work *Systema Naturae*, he also invented the still-used taxonomical naming system called binomial nomenclature (which gives living things two names, such as *Homo sapiens*). Linnaeus believed that species (and organisms) are fixed and unchanging.

John Lubbock (1834–1913) was a mathematician, scientist, banker, politician, and the 1st Baron of Avebury. Like Darwin, he also lived in the village of Down in Kent and became one of Darwin's closest younger friends.

Charles Lyell (1797–1875) was a Scottish geologist and advocate of uniformitarianism: the idea that the same forces that change the physical earth today also operated in the past. Lyell was a professional and personal supporter of Darwin's ideas.

Thomas Robert Malthus (1766–1834) was an English demographer and cleric who theorized that human populations are checked by disease and famine. Both Darwin and Wallace developed their respective theories of natural selection after reading Malthus, whose *An Essay on the Principle of Population* was first published in 1798.

Harriet Martineau (1802–76) was a social theorist and prominent Whig who wrote over 35 notable books on social theory. She was also romantically involved with Darwin's brother Erasmus.

Ernst Mayr (1904–2005) was a German biologist and Nobel Prize winner who, as well as developing the biological species concept, is considered one of the founders of the modern evolutionary synthesis.

Gregor Mendel (1822–84) was a Moravian monk whose 1860s writings on genetics and inheritance–based mainly on his work breeding pea plants and widely unknown until 1900–were one of the "missing pieces" of modern evolutionary theory.

Geoffrey Miller (b. 1965) is an evolutionary psychologist working at the University of New Mexico. His research interests include the study of mate choice in humans.

St George Jackson Mivart (1827-1900) was a British biologist. He is best known for his staunch criticism of natural selection in his 1871 book *On the Genesis of Species*.

G. E. Moore (1873–1958) was an influential British realist philosopher whose systematic approach to ethical problems and remarkably meticulous approach helped found the "analytic" tradition in philosophy. Moore is best known for his defense of ethical non-naturalism and naturalistic philosophy.

Lewis Henry Morgan (1818–81) was an American anthropologist and a contemporary of Darwin whose theories on the technological "evolution" of various societies became widely circulated for several generations, until challenged by more relativist anthropologists.

Thomas Hunt Morgan (1866-1945) was an American geneticist and Nobel Prize winner in Medicine. Morgan was the first to link heredity traits to chromosomes through his studies of fruit flies.

Nettie Stevens (1861-1912) was an American geneticist who worked at Bryn Mawr and at the Carnegie Institute of Washington. She was one of the earliest American women to be recognised for her contributions to science, and is best known for her work on sex determination – in particular, her discovery that the presence or absence of the Y chromosome decides the sex of offspring.

Isaac Newton (1642–1726) was an English physicist who discovered the principles of gravity and laws of motion.

Friedrich Nietzsche (1844–1900) was a German philosopher noted for philosophical concepts such as the Superman and for his contentious work on the usefulness and purpose of religion.

Richard Owen (1804–92) was an English paleontologist and comparative anatomist probably best known for coining the word *Dinosauria*. He was an outspoken opponent of Darwin's theory of evolution by natural selection—although he agreed that evolution occurred, he believed it was driven by a more complex mechanism than that outlined by Darwin.

Bishop William Paley (1743–1805) was an English clergyman and philosopher. Using the now famous watchmaker analogy, he argued in his book *Natural Theology or Evidences of the Existence and Attributes of the Deity* that the existence of God is proved by the beauty and complexity of the natural world.

Steven Pinker (b. 1954) is a Canadian American psychologist well known for his book *The Blank Slate: The Modern Denial of Human Nature*, which takes a biologically determinist stance to human behavior (that is, he believes that our behavior is determined by our biology).

Plato (428–348 BCE) was a classical Greek philosopher who focused primarily on topics to do with justice, beauty, and equality. He founded the Academy in Athens, the first institution of higher learning in the

Western world.

Jean-Jacques Rousseau (1712–78) was a Swiss writer and philosopher. He wrote *The Social Contract* (1762), which challenged the supremacy of the state—or religious authority—over that of the individual.

Étienne Geoffroy Saint-Hilaire (1772–1844) was a French naturalist. He established the principle of unity of composition, suggesting that a single consistent basic plan could be found in all animals.

Adam Sedgwick (1785–1873) was a renowned English geologist and Darwin's former geology teacher at Cambridge, who opposed his pupil's theory of evolution.

John Maynard Smith (1920–2004) was an English evolutionary biologist best known for his work on game theory in an evolutionary context.

Herbert Spencer (1820–1903) was an English biologist and anthropologist. A contemporary of Darwin's, he is also known for coining the term "survival of the fittest."

Walter Sutton (1877–1916) was an American geneticist. Using the work of Gregor Mendel as a starting point, he developed an important theory about chromosomes.

Frederick Temple (1821–1902) was an English clergyman who became Archbishop of Canterbury. He was known for his interest in the way that religion interacts with science.

Michael Tomasello (b.1950) is an American-born psychologist, now based in Germany. His research centers on investigating the origin of human intelligence.

Alfred Russel Wallace (1823–1913) was a Welsh naturalist. Wallace independently came up with a theory of natural selection that was near-identical to Darwin's as the mechanism for species change. Both Darwin and Wallace are recognized as co-discoverers of the concept of natural selection.

Bruce Waltke (b. 1930) is an American Reformed Evangelical professor, best known for his controversial work on theistic evolution. This theory attempts to unite evolution with the Abrahamic belief that God created the world.

Josiah Wedgwood (1730–95) was an English pottery designer and manufacturer. He used a scientific approach to pottery making and was known for his exhaustive research into materials.

Friedrich Leopold August Weismann (1834–1914) was a German evolutionary biologist. He is best known for coming up with the germ plasm theory, which states that (in multicellular organisms) inheritance only takes place by means of the germ cells—the gametes or egg and sperm cells. His idea is vital to the modern evolutionary synthesis.

Edward Osborne Wilson (b. 1929) is an American biologist and environmental activist considered to be the leading thinker and theorist of sociobiology (inquiry into the biological nature of social behavior).

Merlene Zuk (b. 1956) is an American biologist whose research

WORKS CITED

WORKS CITED

Andersson, Malte B. *Sexual Selection*: *Monographs in Behavior and Ecology*. Princeton: Princeton University Press, 1994.

Baden-Powell, Rev. *Philosophy of Creation*. In *Evolution and Dogma* by John Augustine Zahm. Hard Press, 2013.

Baguñà, Jaume, and Jordi Garcia-Fernàndez. "*Evo-Devo*: the Long and Winding Road." *The International Journal of Developmental Biology* 47, nos. 7–8 (2003): 705–13.

Barton, Nicholas H., Derek E. G. Briggs, Jonathan A. Eisen, David B. Goldstein and Nipam H. Patel. *Evolution*. Cold Spring Harbor, NY: Cold Spring Harbor Laboratory Press, 2007.

Bell, Charles. *Essays on the Anatomy and Philosophy of Expression*. Montana: Kessinger Publishing, 2008.

Bowler, Peter. *Charles Darwin: The Man and His Influence*. Cambridge: Cambridge University Press, 1996.

Bräuer, Juliane, Josep Call, and Michael Tomasello. "Chimpanzees really know what others can see in a competitive situation." *Animal Cognition* 10 (2007): 439–48.

Browne, Janet. *Charles Darwin: A Biography. Vol. 1: Voyaging*. Princeton, NJ: Princeton University Press, 1996.

Charles Darwin: A Biography. Vol. 2: The Power of Place. London: Pimlico, 2003.

Buffon, Georges-Louis Leclerc, Comte de. *Les Epoques de la Nature*. In *Histoire Naturelle, générale et particulière, avec la description du Cabinet du Roi*. Paris: imprimerie nationale, 1749–88.

Bulmer, Michael. "The theory of natural selection of Alfred Russel Wallace FRS." *Royal Society Journal of the History of Science* 59, no. 2 (2005): 125–36.

Cadbury, Deborah. *Terrible Lizard: The First Dinosaur Hunters and the Birth of a New Science*. New York: Henry Holt, 2000.

Cañestro, Cristian, Hayato Yokoi, and John H. Postlethwait. "Evolutionary Developmental Biology and Genomics." *Nature Reviews Genetics* 8, no. 12 (2007): 932–42.

Carothers, B. J., and H. T. Reis. "Men and Women Are From Earth: Examining the Latent Structure of Gender." *Journal of Personality and Social Psychology* 104, no. 2 (2013): 385–407.

Chambers, Robert. *Vestiges of the Natural History of Creation*. London: John Churchill, 1844.

Cook, Peter. *Evolution Versus Intelligent Design: Why All the Fuss? The Arguments for Both Sides*. Australia: New Holland Publishing, 2007.

Corning, Peter. *Holistic Darwinism: Synergy, Cybernetics, and the Bioeconomics of Evolution*. Chicago: University of Chicago Press, 2010.

Cuvier, G. *Tableau elementaire de l'histoire naturelle des animaux*. Paris: Baudouin, 1798. Available at: https://archive.org/details/tableaulment00cuvi. [Accessed February 16, 2016.]

Darwin, Charles. *The Descent of Man, and Selection in Relation to Sex*. London: John Murray, 1871.

The Expression of the Emotions in Man and Animals. London: John Murray, 1872.

On the Origin of Species by Means of Natural Selection, or the Preservation of Favoured Races in the Struggle for Life. Introduction and notes by Gillian Beer. Oxford: Oxford University Press, 2008.

The Variation of Animals and Plants Under Domestication. London: John Murray, 1868.

Darwin, Charles, and Alfred Russel Wallace. "On the Tendency of Species to form Varieties; and on the Perpetuation of Varieties and Species by Natural Means of Selection." *Journal of the Proceedings of the Linnean Society of London. Zoology* 3 (20 August, 1858): 45–50.

Dawkins, Richard. "Science, Delusion and the Appetite for Wonder." *Reports of the National Center for Science Education* 17, no. 1 (1997): 8–14.

The Selfish Gene. Oxford: Oxford University Press, 1990.

Delgado, Cynthia. "Finding Evolution in Medicine." *National Institutes of Health Record* 58, no. 15 (2006).

Desmond, Adrian, and James Moore. *Darwin*. London: Michael Joseph, 1991.

Durinx, Michel, and Tom J. M. Van Dooren. "Assortative Mate Choice and Dominance Modification: Alternative Ways of Removing Heterozygote Disadvantage." *Evolution* 63, no. 2 (2009): 334–52.

Economist. "Evolution and Religion: In the Beginning." April 19, 2007. Available at: http://www.economist.com/node/9036706. [Accessed February 4, 2016.]

Eimer, Theodor. *Organic Evolution as the Result of the Inheritance of Acquired Characteristics According to the Laws of Organic Growth*. London: Macmillan, 1890.

Eiseley, Loren. *Darwin's Century*. New York: Anchor Books/Doubleday, 1961.

Elstein, Daniel. "Species as a Social Construction: Is Species Morally Relevant?" *Journal for Critical Animal Studies* 1, no. 1 (2003): 53–71.

Green, Nathan. "Richard Dawkins calls for evolution to be taught to children from age five." *Guardian*, September 1, 2011.

Goodwin, Brian. *How the Leopard Changed Its Spots: The Evolution of Complexity*. Princeton, NJ: Princeton University Press, 2001.

Gould, S. J. "Impeaching a Self-Appointed Judge." *Scientific American* 267, no. 1 (1992): 118–21.

Gould, S. J., and Niles Eldredge. "Punctuated Equilibria: an Alternative to Phyletic Gradualism." In *Models in Paleobiology*, ed. T. J. M. Schopf, 82–115. San Francisco: Freeman Cooper, 1972.

Guldberg, Helene. "Restating the Case for Human Uniqueness." *Psychology Today*, November 8, 2010. Available at: https://www.psychologytoday.com/blog/reclaiming-childhood/201011/restating-the-case-human-uniqueness. [Accessed February 4, 2016.]

Hager, Thomas. *The Life of Linus Pauling*. New York: Simon and Schuster, 1995.

Harris, Mark. "Human uniqueness, and are humans the pinnacle of evolution?" *Science and Religion @ Edinburgh*, September 7, 2014. Available at: http://www.blogs.hss.ed.ac.uk/science-and-religion/2014/09/07/human-uniqueness-and-are-humans-the-pinnacle-of-evolution/. [Accessed February 16, 2016.]

Haviland, W. A., and G.W. Crawford. *Human Evolution and Prehistory*. Cambridge, MA: Harvard University Press, 2002.

Hawkins, Mike. *Social Darwinism in European and American thought, 1860–1945: Nature as Model and Nature as Threat*. Cambridge: Cambridge University Press, 1998.

Hey, Jody. *Genes, Categories and Species*. New York: Oxford University Press 2001.

Hollingdale, R. J. *Nietzsche: The Man and His Philosophy*. Cambridge: Cambridge University Press, 1999.

Hooke, Samuel Henry. *Middle Eastern Mythology*. Dover Publications, 2013.

Hunt, Sonia Y. "Controversies in Treatment Approaches: Gene Therapy, IVF, Stem Cells, and Pharmacogenomics." *Nature Education* 1, no. 1 (2008): 222.

Huxley, Thomas Henry. *Huxley Papers*. London: Imperial College of Science and Technology.

Hyers, Conrad. *The Meaning of Creation: Genesis and Modern Science.* Louisville: Westminster John Knox, 1984.

Joel, Daphna, Zohar Berman, Ido Tavor, Nadav Wexler, Olga Gaber, Yaniv Stein, Nisan Shefi, Jared Pool, Sebastian Urchs, Daniel S. Margulies, Franziskus Liem, Jürgen Hänggi, Lutz Jäncke, and Yaniv Assaf. "Sex beyond the genitalia: the human brain mosaic." *PNAS* 112, no. 50 (2015):15468–73.

Johnson, Phillip E. "The Church of Darwin." *Wall Street Journal,* August 16, 1999.

Jones, Steve. *Darwin's Island: The Galapagos in the Garden of England.* London: Little Brown, 2009.

Kitcher, Philip. *Living with Darwin: Evolution, Design, and the Future of Faith.* New York; Oxford: Oxford University Press, 2007.

Krebs, J. R., and A. Kacelnik. *Behavioural Ecology: An Evolutionary Approach.* Oxford: Blackwell Scientific, 1991.

Johnson, Phillip E. *Darwin on Trial.* Downers Grove, IL: InterVarsity Press, 1991.

Lamarck, J. B. *Zoological Philosophy: An Exposition with Regard to the Natural History of Animals.* Translated by Hugh Elliot. Chicago: University of Chicago Press, 1984.

Larson, Edward J. E. *Evolution: The Remarkable History of a Scientific Theory.* New York: Modern Library, 2004.

Leff, David. "About Charles Darwin." Available at: www.aboutDarwin.com. [Accessed February 4, 2016.]

Levine, George. *Darwin the Writer.* Oxford: Oxford University Press, 2011.

Lewontin, R. C. "Biological Determinism." The Tanner Lectures on Human Values, University of Utah, March 31 and April 1, 1982. Available at: http://tannerlectures.utah.edu/_documents/a-to-z/l/lewontin83.pdf. [Accessed February 5, 2016.]

Lyell, Charles. *The Principles of Geology: Being an Attempt to Explain the Former Changes of the Earth's Surface, by Reference to Causes Now in Operation.* 3 Volumes. London: John Murray, 1830–3.

Magner, Lois N. *A History of the Life Sciences.* New York; Basel: Marcel Dekker, 1994.

Mallet, J. A. "Species definition for the modern synthesis." *Trends in Ecology and Evolution* 10 (1995): 294–9

Mayr, Ernst. "Darwin's Influence on Modern Thought." *Proceedings of the American Philosophical Society* 139, no. 4 (Dec 1995): 317–25.

Mendel, Gregor. "Versuche über Pflanzenhybriden." In *Verhandlungen des naturforschenden Vereins in Brünn*, 1866.

Miller, David. "Natural Selection and its Scientific Status." *Popper Selections*, ed. David Miller. Princeton, NJ: Princeton University Press, 1985.

Moore, G. E. *Principia Ethica*. Cambridge: Cambridge University Press, 1993.

Morgan, Thomas Hunt, A. H. Sturtevant, H. J. Muller, and C. B. Bridges. *The Mechanism of Mendelian Heredity*. New York: Henry Holt and Company, 1915.

Morris, Jeffrey J., Richard E. Lenski, and Erik R. Zinser. "The Black Queen Hypothesis: Evolution of Dependencies Through Adaptive Gene Loss." *MBio* 3, no. 2 (2012): 1–7.

Nowak, Martin A. "Five rules for the evolution of cooperation." *Science* 314 (2006): 1560–3.

Ollerton, J. "Speciation: Flowering time and the Wallace Effect." *Heredity* 95, no. 3 (2005): 181–2

Orr, H. Allen. "Testing Natural Selection with Genetics." *Scientific American* 300, no. 1 (2009): 44.

Paley, William. *Natural Theology: or, Evidences of the Existence and Attributes of the Deity*. London: J. Faulder, 1809.

Pew Forum on Religion and Public Life. "Religious Groups: Opinions of Evolution." February 4, 2009. Available at: http://www.pewforum. org/2009/02/04/religious-differences-on-the-question-of-evolution/. [Accessed February 5, 2016.]

Popper, Karl. "Natural Selection and its Scientific Status." *Popper Selections*, ed. David Miller, 241–43. Princeton, NJ: Princeton University Press, 1985.

Reed, Edward S. "The Lawfulness of Natural Selection." *American Naturalist* 118, no. 1 (1981): 61–71.

Ridley, Mark. *How to Read Darwin*. London: Granta Books, 2006.

Ridley, Matt. "Darwin's Legacy: Modern Darwins." *National Geographic*, February 2009.

Robinson, Bruce A. "U.S. public opinion polls about evolution & creation." Available at: http://www.religioustolerance.org/ev_publi2007.htm. [Accessed February 16, 2016.]

Rutter, Michael. *Genes and Behavior: Nature-Nurture Interplay Explained*. Oxford: Blackwell, 2006.

Sachs, Joel L. "Cooperation Within and Among species." *Journal of Evolutionary Biology* 19, no. 5 (2006): 1415–8.

Secord, James A. *Victorian Sensation: The Extraordinary Publication, Reception, and Secret Authorship of Vestiges of the Natural History of Creation*. Chicago: University of Chicago Press, 2000.

Sedgwick, Adam. "Review of Vestiges." *Edinburgh Review* 82 (July 1845): 1–85.

Sewell, Dennis. *The Political Gene: How Darwin's Ideas Changed Politics*. London: Picador, 2009.

Slevin, Peter. "Battle on Teaching Evolution Sharpens." *Washington Post*, March 14, 2005.

Smith, Charles. H. "Wallace's Unfinished Business: The 'Other Man' in Evolutionary Theory." *Complexity* 10, no 2 (2004): 25–32.

Stott, Rebecca. *Darwin's Ghosts: In Search of the First Evolutionists*. London: Bloomsbury, 2012.

Strobel, Lee. *The Case for a Creator: A Journalist Investigates Scientific Evidence That Points Toward God*. Grand Rapids, MI: Zondervan, 2004.

Tanner, Julia. "The Naturalistic Fallacy." *Richmond Journal of Philosophy* 13 (2006): 1–6.

Tomasello, Michael, Malinda Carpenter, Josep Call, Tanya Behne, and Henrike Moll. "Understanding and sharing intentions: the origins of cultural cognition." *Behavioral and Brain Sciences* 28, no. 5 (2005): 675–735.

Wallace, Alfred Russel. *Darwinism: An Exposition of the Theory of Natural Selection, with Some of Its Applications.* London: Macmillan & Co, 1889.

"Letter to George Silk." Natural History Museum. Wallace Letters Online: Letter WCP373.373.

My Life: A Record of Events and Opinions. London: Chapman and Hall, 1905.

Weikart, R. *From Darwin to Hitler: Evolutionary Ethics, Eugenics and Racism in Germany.* London: Palgrave Macmillan; 2004.

Weismann, Friedrich Leopold August. *Die Entstehung der Sexualzellen bei den Hydromedusen: Zugleich ein Beitrag zur Kenntniss des Baues und der Lebenserscheinungen dieser Gruppe*. Jena: Fischer, 1883.

Wildman, Derek. E., Monica Uddin, Guozhen Liu, Lawrence I. Grossman, and Morris Goodman. "Implications of natural selection in shaping 99.4% nonsynonymous DNA identity between humans and chimpanzees: Enlarging genus Homo." *Proceedings of the National Academy of Sciences of the United States of America (PNAS)* 100, no. 12 (2003): 7181–8.

Wilkins, John. *Species: A History of the Idea.* Oakland, CA: University of California Press, 2011.

Williams, George C. "Pleiotropy, natural selection, and the evolution of senescence," *Evolution* 11, no. 4 (1957): 398–411.

Wilson, E. O. "Human Decency Is Animal." *New York Times Magazine*, October 12, 1975.

Witham, Larry. "Many Scientists See God's Hand in Evolution." *Reports of the National Center for Science Education* 17, no. 6 (November–December 1997): 33.

Wyhe, J. van. "Mind the Gap: Did Darwin Avoid Publishing His Theory For Many Years?" *Royal Society Journal of the History of Science* 61, no. 2 (2007): 177–205.

Wynne-Edwards, V. C. *Evolution Through Group Selection*. Oxford: Blackwell Scientific, 2006.

THE MACAT LIBRARY
BY DISCIPLINE

AFRICANA STUDIES

Chinua Achebe's *An Image of Africa: Racism in Conrad's Heart of Darkness*
W. E. B. Du Bois's *The Souls of Black Folk*
Zora Neale Huston's *Characteristics of Negro Expression*
Martin Luther King Jr's *Why We Can't Wait*
Toni Morrison's *Playing in the Dark: Whiteness in the American Literary Imagination*

ANTHROPOLOGY

Arjun Appadurai's *Modernity at Large: Cultural Dimensions of Globalisation*
Philippe Ariès's *Centuries of Childhood*
Franz Boas's *Race, Language and Culture*
Kim Chan & Renée Mauborgne's *Blue Ocean Strategy*
Jared Diamond's *Guns, Germs & Steel: the Fate of Human Societies*
Jared Diamond's *Collapse: How Societies Choose to Fail or Survive*
E. E. Evans-Pritchard's *Witchcraft, Oracles and Magic Among the Azande*
James Ferguson's *The Anti-Politics Machine*
Clifford Geertz's *The Interpretation of Cultures*
David Graeber's *Debt: the First 5000 Years*
Karen Ho's *Liquidated: An Ethnography of Wall Street*
Geert Hofstede's *Culture's Consequences: Comparing Values, Behaviors, Institutes and Organizations across Nations*
Claude Lévi-Strauss's *Structural Anthropology*
Jay Macleod's *Ain't No Makin' It: Aspirations and Attainment in a Low-Income Neighborhood*
Saba Mahmood's *The Politics of Piety: The Islamic Revival and the Feminist Subject*
Marcel Mauss's *The Gift*

BUSINESS

Jean Lave & Etienne Wenger's *Situated Learning*
Theodore Levitt's *Marketing Myopia*
Burton G. Malkiel's *A Random Walk Down Wall Street*
Douglas McGregor's *The Human Side of Enterprise*
Michael Porter's *Competitive Strategy: Creating and Sustaining Superior Performance*
John Kotter's *Leading Change*
C. K. Prahalad & Gary Hamel's *The Core Competence of the Corporation*

CRIMINOLOGY

Michelle Alexander's *The New Jim Crow: Mass Incarceration in the Age of Colorblindness*
Michael R. Gottfredson & Travis Hirschi's *A General Theory of Crime*
Richard Herrnstein & Charles A. Murray's *The Bell Curve: Intelligence and Class Structure in American Life*
Elizabeth Loftus's *Eyewitness Testimony*
Jay Macleod's *Ain't No Makin' It: Aspirations and Attainment in a Low-Income Neighborhood*
Philip Zimbardo's *The Lucifer Effect*

ECONOMICS

Janet Abu-Lughod's *Before European Hegemony*
Ha-Joon Chang's *Kicking Away the Ladder*
David Brion Davis's *The Problem of Slavery in the Age of Revolution*
Milton Friedman's *The Role of Monetary Policy*
Milton Friedman's *Capitalism and Freedom*
David Graeber's *Debt: the First 5000 Years*
Friedrich Hayek's *The Road to Serfdom*
Karen Ho's *Liquidated: An Ethnography of Wall Street*

John Maynard Keynes's *The General Theory of Employment, Interest and Money*
Charles P. Kindleberger's *Manias, Panics and Crashes*
Robert Lucas's *Why Doesn't Capital Flow from Rich to Poor Countries?*
Burton G. Malkiel's *A Random Walk Down Wall Street*
Thomas Robert Malthus's *An Essay on the Principle of Population*
Karl Marx's *Capital*
Thomas Piketty's *Capital in the Twenty-First Century*
Amartya Sen's *Development as Freedom*
Adam Smith's *The Wealth of Nations*
Nassim Nicholas Taleb's *The Black Swan: The Impact of the Highly Improbable*
Amos Tversky's & Daniel Kahneman's *Judgment under Uncertainty: Heuristics and Biases*
Mahbub Ul Haq's *Reflections on Human Development*
Max Weber's *The Protestant Ethic and the Spirit of Capitalism*

FEMINISM AND GENDER STUDIES

Judith Butler's *Gender Trouble*
Simone De Beauvoir's *The Second Sex*
Michel Foucault's *History of Sexuality*
Betty Friedan's *The Feminine Mystique*
Saba Mahmood's *The Politics of Piety: The Islamic Revival and the Feminist Subjec*t
Joan Wallach Scott's *Gender and the Politics of History*
Mary Wollstonecraft's *A Vindication of the Rights of Woman*
Virginia Woolf's *A Room of One's Own*

GEOGRAPHY

The Brundtland Report's *Our Common Future*
Rachel Carson's *Silent Spring*
Charles Darwin's *On the Origin of Species*
James Ferguson's *The Anti-Politics Machine*
Jane Jacobs's *The Death and Life of Great American Cities*
James Lovelock's *Gaia: A New Look at Life on Earth*
Amartya Sen's *Development as Freedom*
Mathis Wackernagel & William Rees's *Our Ecological Footprint*

HISTORY

Janet Abu-Lughod's *Before European Hegemony*
Benedict Anderson's *Imagined Communities*
Bernard Bailyn's *The Ideological Origins of the American Revolution*
Hanna Batatu's *The Old Social Classes And The Revolutionary Movements Of Iraq*
Christopher Browning's *Ordinary Men: Reserve Police Batallion 101 and the Final Solution in Poland*
Edmund Burke's *Reflections on the Revolution in France*
William Cronon's *Nature's Metropolis: Chicago And The Great West*
Alfred W. Crosby's *The Columbian Exchange*
Hamid Dabashi's *Iran: A People Interrupted*
David Brion Davis's *The Problem of Slavery in the Age of Revolution*
Nathalie Zemon Davis's *The Return of Martin Guerre*
Jared Diamond's *Guns, Germs & Steel: the Fate of Human Societies*
Frank Dikotter's *Mao's Great Famine*
John W Dower's *War Without Mercy: Race And Power In The Pacific War*
W. E. B. Du Bois's *The Souls of Black Folk*
Richard J. Evans's *In Defence of History*
Lucien Febvre's *The Problem of Unbelief in the 16th Century*
Sheila Fitzpatrick's *Everyday Stalinism*

Eric Foner's *Reconstruction: America's Unfinished Revolution, 1863-1877*
Michel Foucault's *Discipline and Punish*
Michel Foucault's *History of Sexuality*
Francis Fukuyama's *The End of History and the Last Man*
John Lewis Gaddis's *We Now Know: Rethinking Cold War History*
Ernest Gellner's *Nations and Nationalism*
Eugene Genovese's *Roll, Jordan, Roll: The World the Slaves Made*
Carlo Ginzburg's *The Night Battles*
Daniel Goldhagen's *Hitler's Willing Executioners*
Jack Goldstone's *Revolution and Rebellion in the Early Modern World*
Antonio Gramsci's *The Prison Notebooks*
Alexander Hamilton, John Jay & James Madison's *The Federalist Papers*
Christopher Hill's *The World Turned Upside Down*
Carole Hillenbrand's *The Crusades: Islamic Perspectives*
Thomas Hobbes's *Leviathan*
Eric Hobsbawm's *The Age Of Revolution*
John A. Hobson's *Imperialism: A Study*
Albert Hourani's *History of the Arab Peoples*
Samuel P. Huntington's *The Clash of Civilizations and the Remaking of World Order*
C. L. R. James's *The Black Jacobins*
Tony Judt's *Postwar: A History of Europe Since 1945*
Ernst Kantorowicz's *The King's Two Bodies: A Study in Medieval Political Theology*
Paul Kennedy's *The Rise and Fall of the Great Powers*
Ian Kershaw's *The "Hitler Myth": Image and Reality in the Third Reich*
John Maynard Keynes's *The General Theory of Employment, Interest and Money*
Charles P. Kindleberger's *Manias, Panics and Crashes*
Martin Luther King Jr's *Why We Can't Wait*
Henry Kissinger's *World Order: Reflections on the Character of Nations and the Course of History*
Thomas Kuhn's *The Structure of Scientific Revolutions*
Georges Lefebvre's *The Coming of the French Revolution*
John Locke's *Two Treatises of Government*
Niccolò Machiavelli's *The Prince*
Thomas Robert Malthus's *An Essay on the Principle of Population*
Mahmood Mamdani's *Citizen and Subject: Contemporary Africa And The Legacy Of Late Colonialism*
Karl Marx's *Capital*
Stanley Milgram's *Obedience to Authority*
John Stuart Mill's *On Liberty*
Thomas Paine's *Common Sense*
Thomas Paine's *Rights of Man*
Geoffrey Parker's *Global Crisis: War, Climate Change and Catastrophe in the Seventeenth Century*
Jonathan Riley-Smith's *The First Crusade and the Idea of Crusading*
Jean-Jacques Rousseau's *The Social Contract*
Joan Wallach Scott's *Gender and the Politics of History*
Theda Skocpol's *States and Social Revolutions*
Adam Smith's *The Wealth of Nations*
Timothy Snyder's *Bloodlands: Europe Between Hitler and Stalin*
Sun Tzu's *The Art of War*
Keith Thomas's *Religion and the Decline of Magic*
Thucydides's *The History of the Peloponnesian War*
Frederick Jackson Turner's *The Significance of the Frontier in American History*
Odd Arne Westad's *The Global Cold War: Third World Interventions And The Making Of Our Times*

LITERATURE

Chinua Achebe's *An Image of Africa: Racism in Conrad's Heart of Darkness*
Roland Barthes's *Mythologies*
Homi K. Bhabha's *The Location of Culture*
Judith Butler's *Gender Trouble*
Simone De Beauvoir's *The Second Sex*
Ferdinand De Saussure's *Course in General Linguistics*
T. S. Eliot's *The Sacred Wood: Essays on Poetry and Criticism*
Zora Neale Huston's *Characteristics of Negro Expression*
Toni Morrison's *Playing in the Dark: Whiteness in the American Literary Imagination*
Edward Said's *Orientalism*
Gayatri Chakravorty Spivak's *Can the Subaltern Speak?*
Mary Wollstonecraft's *A Vindication of the Rights of Women*
Virginia Woolf's *A Room of One's Own*

PHILOSOPHY

Elizabeth Anscombe's *Modern Moral Philosophy*
Hannah Arendt's *The Human Condition*
Aristotle's *Metaphysics*
Aristotle's *Nicomachean Ethics*
Edmund Gettier's *Is Justified True Belief Knowledge?*
Georg Wilhelm Friedrich Hegel's *Phenomenology of Spirit*
David Hume's *Dialogues Concerning Natural Religion*
David Hume's *The Enquiry for Human Understanding*
Immanuel Kant's *Religion within the Boundaries of Mere Reason*
Immanuel Kant's *Critique of Pure Reason*
Søren Kierkegaard's *The Sickness Unto Death*
Søren Kierkegaard's *Fear and Trembling*
C. S. Lewis's *The Abolition of Man*
Alasdair MacIntyre's *After Virtue*
Marcus Aurelius's *Meditations*
Friedrich Nietzsche's *On the Genealogy of Morality*
Friedrich Nietzsche's *Beyond Good and Evil*
Plato's *Republic*
Plato's *Symposium*
Jean-Jacques Rousseau's *The Social Contract*
Gilbert Ryle's *The Concept of Mind*
Baruch Spinoza's *Ethics*
Sun Tzu's *The Art of War*
Ludwig Wittgenstein's *Philosophical Investigations*

POLITICS

Benedict Anderson's *Imagined Communities*
Aristotle's *Politics*
Bernard Bailyn's *The Ideological Origins of the American Revolution*
Edmund Burke's *Reflections on the Revolution in France*
John C. Calhoun's *A Disquisition on Government*
Ha-Joon Chang's *Kicking Away the Ladder*
Hamid Dabashi's *Iran: A People Interrupted*
Hamid Dabashi's *Theology of Discontent: The Ideological Foundation of the Islamic Revolution in Iran*
Robert Dahl's *Democracy and its Critics*
Robert Dahl's *Who Governs?*
David Brion Davis's *The Problem of Slavery in the Age of Revolution*

The Macat Library By Discipline

Alexis De Tocqueville's *Democracy in America*
James Ferguson's *The Anti-Politics Machine*
Frank Dikotter's *Mao's Great Famine*
Sheila Fitzpatrick's *Everyday Stalinism*
Eric Foner's *Reconstruction: America's Unfinished Revolution, 1863-1877*
Milton Friedman's *Capitalism and Freedom*
Francis Fukuyama's *The End of History and the Last Man*
John Lewis Gaddis's *We Now Know: Rethinking Cold War History*
Ernest Gellner's *Nations and Nationalism*
David Graeber's *Debt: the First 5000 Years*
Antonio Gramsci's *The Prison Notebooks*
Alexander Hamilton, John Jay & James Madison's *The Federalist Papers*
Friedrich Hayek's *The Road to Serfdom*
Christopher Hill's *The World Turned Upside Down*
Thomas Hobbes's *Leviathan*
John A. Hobson's *Imperialism: A Study*
Samuel P. Huntington's *The Clash of Civilizations and the Remaking of World Order*
Tony Judt's *Postwar: A History of Europe Since 1945*
David C. Kang's *China Rising: Peace, Power and Order in East Asia*
Paul Kennedy's *The Rise and Fall of Great Powers*
Robert Keohane's *After Hegemony*
Martin Luther King Jr.'s *Why We Can't Wait*
Henry Kissinger's *World Order: Reflections on the Character of Nations and the Course of History*
John Locke's *Two Treatises of Government*
Niccolò Machiavelli's *The Prince*
Thomas Robert Malthus's *An Essay on the Principle of Population*
Mahmood Mamdani's *Citizen and Subject: Contemporary Africa And The Legacy Of Late Colonialism*
Karl Marx's *Capital*
John Stuart Mill's *On Liberty*
John Stuart Mill's *Utilitarianism*
Hans Morgenthau's *Politics Among Nations*
Thomas Paine's *Common Sense*
Thomas Paine's *Rights of Man*
Thomas Piketty's *Capital in the Twenty-First Century*
Robert D. Putman's *Bowling Alone*
John Rawls's *Theory of Justice*
Jean-Jacques Rousseau's *The Social Contract*
Theda Skocpol's *States and Social Revolutions*
Adam Smith's *The Wealth of Nations*
Sun Tzu's *The Art of War*
Henry David Thoreau's *Civil Disobedience*
Thucydides's *The History of the Peloponnesian War*
Kenneth Waltz's *Theory of International Politics*
Max Weber's *Politics as a Vocation*
Odd Arne Westad's *The Global Cold War: Third World Interventions And The Making Of Our Times*

POSTCOLONIAL STUDIES

Roland Barthes's *Mythologies*
Frantz Fanon's *Black Skin, White Masks*
Homi K. Bhabha's *The Location of Culture*
Gustavo Gutiérrez's *A Theology of Liberation*
Edward Said's *Orientalism*
Gayatri Chakravorty Spivak's *Can the Subaltern Speak?*

PSYCHOLOGY

Gordon Allport's *The Nature of Prejudice*
Alan Baddeley & Graham Hitch's *Aggression: A Social Learning Analysis*
Albert Bandura's *Aggression: A Social Learning Analysis*
Leon Festinger's *A Theory of Cognitive Dissonance*
Sigmund Freud's *The Interpretation of Dreams*
Betty Friedan's *The Feminine Mystique*
Michael R. Gottfredson & Travis Hirschi's *A General Theory of Crime*
Eric Hoffer's *The True Believer: Thoughts on the Nature of Mass Movements*
William James's *Principles of Psychology*
Elizabeth Loftus's *Eyewitness Testimony*
A. H. Maslow's *A Theory of Human Motivation*
Stanley Milgram's *Obedience to Authority*
Steven Pinker's *The Better Angels of Our Nature*
Oliver Sacks's *The Man Who Mistook His Wife For a Hat*
Richard Thaler & Cass Sunstein's *Nudge: Improving Decisions About Health, Wealth and Happiness*
Amos Tversky's *Judgment under Uncertainty: Heuristics and Biases*
Philip Zimbardo's *The Lucifer Effect*

SCIENCE

Rachel Carson's *Silent Spring*
William Cronon's *Nature's Metropolis: Chicago And The Great West*
Alfred W. Crosby's *The Columbian Exchange*
Charles Darwin's *On the Origin of Species*
Richard Dawkin's *The Selfish Gene*
Thomas Kuhn's *The Structure of Scientific Revolutions*
Geoffrey Parker's *Global Crisis: War, Climate Change and Catastrophe in the Seventeenth Century*
Mathis Wackernagel & William Rees's *Our Ecological Footprint*

SOCIOLOGY

Michelle Alexander's *The New Jim Crow: Mass Incarceration in the Age of Colorblindness*
Gordon Allport's *The Nature of Prejudice*
Albert Bandura's *Aggression: A Social Learning Analysis*
Hanna Batatu's *The Old Social Classes And The Revolutionary Movements Of Iraq*
Ha-Joon Chang's *Kicking Away the Ladder*
W. E. B. Du Bois's *The Souls of Black Folk*
Émile Durkheim's *On Suicide*
Frantz Fanon's *Black Skin, White Masks*
Frantz Fanon's *The Wretched of the Earth*
Eric Foner's *Reconstruction: America's Unfinished Revolution, 1863-1877*
Eugene Genovese's *Roll, Jordan, Roll: The World the Slaves Made*
Jack Goldstone's *Revolution and Rebellion in the Early Modern World*
Antonio Gramsci's *The Prison Notebooks*
Richard Herrnstein & Charles A Murray's *The Bell Curve: Intelligence and Class Structure in American Life*
Eric Hoffer's *The True Believer: Thoughts on the Nature of Mass Movements*
Jane Jacobs's *The Death and Life of Great American Cities*
Robert Lucas's *Why Doesn't Capital Flow from Rich to Poor Countries?*
Jay Macleod's *Ain't No Makin' It: Aspirations and Attainment in a Low Income Neighborhood*
Elaine May's *Homeward Bound: American Families in the Cold War Era*
Douglas McGregor's *The Human Side of Enterprise*
C. Wright Mills's *The Sociological Imagination*

Thomas Piketty's *Capital in the Twenty-First Century*
Robert D. Putman's *Bowling Alone*
David Riesman's *The Lonely Crowd: A Study of the Changing American Character*
Edward Said's *Orientalism*
Joan Wallach Scott's *Gender and the Politics of History*
Theda Skocpol's *States and Social Revolutions*
Max Weber's *The Protestant Ethic and the Spirit of Capitalism*

THEOLOGY

Augustine's *Confessions*
Benedict's *Rule of St Benedict*
Gustavo Gutiérrez's *A Theology of Liberation*
Carole Hillenbrand's *The Crusades: Islamic Perspectives*
David Hume's *Dialogues Concerning Natural Religion*
Immanuel Kant's *Religion within the Boundaries of Mere Reason*
Ernst Kantorowicz's *The King's Two Bodies: A Study in Medieval Political Theology*
Søren Kierkegaard's *The Sickness Unto Death*
C. S. Lewis's *The Abolition of Man*
Saba Mahmood's *The Politics of Piety: The Islamic Revival and the Feminist Subject*
Baruch Spinoza's *Ethics*
Keith Thomas's *Religion and the Decline of Magic*

COMING SOON

Chris Argyris's *The Individual and the Organisation*
Seyla Benhabib's *The Rights of Others*
Walter Benjamin's *The Work Of Art in the Age of Mechanical Reproduction*
John Berger's *Ways of Seeing*
Pierre Bourdieu's *Outline of a Theory of Practice*
Mary Douglas's *Purity and Danger*
Roland Dworkin's *Taking Rights Seriously*
James G. March's *Exploration and Exploitation in Organisational Learning*
Ikujiro Nonaka's *A Dynamic Theory of Organizational Knowledge Creation*
Griselda Pollock's *Vision and Difference*
Amartya Sen's *Inequality Re-Examined*
Susan Sontag's *On Photography*
Yasser Tabbaa's *The Transformation of Islamic Art*
Ludwig von Mises's *Theory of Money and Credit*

Macat Disciplines

Access the greatest ideas and thinkers across entire disciplines, including

FEMINISM, GENDER AND QUEER STUDIES

Simone De Beauvoir's
The Second Sex

Michel Foucault's
History of Sexuality

Betty Friedan's
The Feminine Mystique

Saba Mahmood's
*The Politics of Piety:
The Islamic Revival and
the Feminist Subject*

Joan Wallach Scott's
*Gender and the
Politics of History*

Mary Wollstonecraft's
*A Vindication of the
Rights of Woman*

Virginia Woolf's
A Room of One's Own

Judith Butler's
Gender Trouble

Macat analyses are available from all good bookshops and libraries.

Access hundreds of analyses through one, multimedia tool.
Join free for one month **library.macat.com**

Macat Disciplines

Access the greatest ideas and thinkers across entire disciplines, including

CRIMINOLOGY

Michelle Alexander's
*The New Jim Crow:
Mass Incarceration in the
Age of Colorblindness*

**Michael R. Gottfredson
& Travis Hirschi's**
A General Theory of Crime

Elizabeth Loftus's
Eyewitness Testimony

**Richard Herrnstein
& Charles A. Murray's**
*The Bell Curve: Intelligence and
Class Structure in American Life*

Jay Macleod's
*Ain't No Makin' It:
Aspirations and Attainment in a
Low-Income Neighborhood*

Philip Zimbardo's
The Lucifer Effect

Macat analyses are available from all good bookshops and libraries.

Access hundreds of analyses through one, multimedia tool.
Join free for one month **library.macat.com**

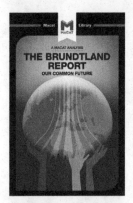

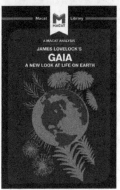
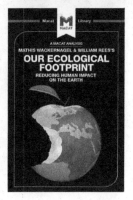